# A VIEW FROM THE TRACTOR

*Wit and Wisdom from the
Nation's Favourite Dairy Farmer*

## Roger Evans

MERLIN UNWIN BOOKS

First published in Great Britain by Merlin Unwin Books, 2014
Reprinted 2014, 2016, 2017, 2019 and 2022

Text © Roger Evans 2014

Published by:
Merlin Unwin Books Ltd
Palmers House, 7 Corve Street
Ludlow, Shropshire SY8 1DB
U.K.
www.merlinunwin.co.uk

The author asserts his moral right to be identified with this work.

Designed and set in Times New Roman 11pt by Merlin Unwin

ISBN 978 1 913159 54 2

For the first time ever we have three bulls. There's Peter the British Blue bull who lives here at home. We have a Limousin bull who always lives at a place we rent a couple of miles away. He's never actually been home, which is quite sad, and because he's not been home and has not had the opportunity to develop the sort of closeness our other animals have, he doesn't have a name, which is even sadder. Then we have a young Friesian bull that I bought as a calf last year at the dispersal sale of a well-known herd in Gloucestershire and whose name is Ben.

I suppose Ben is half-grown now, he lives in a shed on his own and is quite friendly. But there's a very considerable chance that he won't always be friendly. In fact, for reasons of safety, we have to assume that he will be dangerous. So there's two jobs to do with regard to Ben. We have to build somewhere for him to live that is safe. It has to be safe to feed him, to put bedding under him, to clean his pen out and for us to bring cows to him that he can get into calf, so some of the excellent breeding that is his background in Gloucestershire, transmits to our herd. And we have to put a ring through his nose. I don't ever see us leading him about on that ring, but should we ever need to handle him, it's an important piece of control mechanism. It's also a signal to everyone, 'Here is a bull, take care'.

I raised this issue in the pub. The advice was unanimous. 'Don't do it yourselves, he'll always remember who puts the ring in his nose, he'll remember that it hurt like hell, and about four years later when you've forgotten about it and you drop your guard, he'll have you.' I do not know if this is true but it's advice probably worth heeding. The conversation moves on to lambs and lambing, and I do a quick scan around the table to see who I will get to put the ring in his nose. There's three could do it but they are old friends and I like them. There's some 'newcomers' on another table, now there's a thought...

\*\*\*

Two of my grandchildren used to have an old Shetland pony, they never used to ride him because he had 'attitude' but he was a pet they thought a lot of. He had to be put down in the end because he suffered from laminitis on spring grass and they had endless complaints to the RSPCA from ramblers. The pony's opinion was never sought on this issue but I suspect that, given a choice, he would have toughed it out for a month and carried on for another year. My little grandson wasn't too bothered about the pony's demise (it used to bite him) but his sister was. So I bought her a poultry arc and three rare breed pullets. Other people gave her hens out of a battery unit and she built up quite a little business selling eggs in the village.

Meanwhile, we've just had our point-of-lay pullets go out and with nearly 40,000 to go, there are sometimes escapees. This time there were six so I put them in a box and took them down to Katie. I hadn't really taken much notice of her poultry enterprise lately but I was amazed at the difference between her original rare breed hens and the others, which are brown egg-laying hybrids. The three originals are huge in comparison. They have quite a large grassy enclosure and these three reminded me, for some reason, of galleons under full sail as they made their way sedately about. I think they are Buff Orpingtons.

***

The week before Easter, I spent a couple of days putting electric fences up for the cows. It was so hot I took my shirt off. Not a pretty sight but only me and the dog to see it. Electric fencing without your shirt on is not without its dangers. A big armful of fence stakes can pinch your skin. I lost my left nipple years ago, not a big issue; I never really knew what it was for anyway. But your shirt off in the hot sun tops up your vitamin D levels, starts off your tan and as it's still only March, feels quite good. A week later and there's four inches of snow outside. But that's not all that's outside. Where are my overalls, where are my wellies, where's my coat, where's my hat? They are all out and about somewhere, in sheds, on tractors, possibly lost forever, wherever they have been discarded. It's a long cold wet half hour that gets most of them back together.

## APRIL 28TH 2012

I used to have a yearning to keep sheep. There's so much pleasure and pride in it. To drive around inspecting your ewes and lambs in the evenings in the spring. To look at your ewes all penned up in a group the day you put the tups in. To put a pen of 40 lambs in the market and note that they look as good as anyone else's. I miss all of that. I miss being able to contribute to the 'sheepy' conversations in the pub. There's something about a sheep that makes it always test you. It might do that by just getting out of the field where you seek to confine it. It might present you with two fit healthy lambs with ease, but it much prefers two lambs that are all jumbled up together, heads and tails all this way and that, and it much prefers that you should be required to sort this tangle out at three o'clock in the morning. Sheep constantly challenge you and, in a perverse way, overcoming those challenges brings with it much of the satisfaction. The ultimate sheep ambition is to die, so just keeping it alive is part of the challenge. There's a lot of that that I don't miss. I don't miss staying up all night. I don't miss going around with bottles of milk, topping up hungry lambs. So let them talk about sheep as much as they like – anyway, I'll soon change the subject.

\*\*\*

I share with you the things I see when I'm out and about on the tractor. What I see, other people see. Sometimes they leave me feeling envious. Stephen who works here had a day spreading fertiliser and reports back next morning. 'Saw some curlews yesterday.' 'Where?' (You have to drag it out of him.) 'They came up out of the stream.' 'Which stream, how many of them?' Seems he saw a pair of curlews in the stream that borders the field next to the house. I hadn't seen curlews here for years, yet he sees a pair 300 yards from the house! I don't ask him if they got up and made that beautiful call of theirs. I'd rather not know. But there's more to come. 'Those two deer were up the hollow again.' There's been two deer hanging about here for some weeks now. I'm yet to see them either. They pop in and out of the wood to eat my grass if they want to. 'One's got a broken leg.' Deer are the ultimate free-range animal around here, and I suppose all over the UK. They pop over fences with ease. With such ease that they sometimes become careless and drop one hind leg just a bit. The leg catches the barbed wire that is usually found at the top strand of a fence and the foot goes under the next wire down, and snap! I used to have a business that exported game to France and we often used to get deer in that had broken one leg in this manner. In lots of cases the piece at the bottom of the leg had broken off and they had made the best of things going around on the stump.

***

It's Easter Sunday evening and it's freezing. Can't expect much else, there were four inches of snow here on Friday. But it's dried off now and I'm cutting the lawns. The sun has come out giving an illusion of warmth, but it is only an illusion and I am freezing. I should have put my overalls on and a jacket but it looked quite pleasant when I looked through the kitchen window. Overalls and jacket are only 20 yards away in the kitchen but I won't stop to fetch them – it's that bit of me that won't be beaten. Some people call it stubbornness. So I shiver and persevere and am quite pleased when I get to the last patch, ten minutes should do it. And then there's a bang and a puff of smoke and we stop in our tracks.

I'm not much of a mechanic so I don't know what's happened but I know quite a bit about death and this lawnmower is very dead. I walk away and leave it without a backward glance and I'm soon in the welcome warmth of the kitchen. But when everyone's back at work on Tuesday, I have to do something about it because the grass is growing and we need the lawns trim for our bed and breakfast guests. I take it to where we bought it. There's something in their culture that seeks to blame you for what has gone wrong. 'Have you ever tipped it on its side?' 'Yes.' 'Why did you do that?' 'To put new belts on.' 'Ah well that's your trouble, the oil has got on top of the piston and when you try to start it, the oil won't compress and you've broken the rod.' There's a smile of satisfaction creeps in with this explanation: it's the customer's fault! 'That was two years ago.' 'Oh!' That's not so good. 'What do you think is wrong?' he asks. Why ask me, I'm not a lawnmower mechanic. 'It could be that a valve has broken off and smashed the piston,' I venture. Turns out I'm right. 'What you need is a new mower.' A new mower is £2,000, a new engine is £1,000.

There was a time in dairy farming when you would shrug your shoulders and buy a new mower, you wouldn't be happy about it, but you could cope. These days you have to give it some thought. Which is what I do. At home my no. 2 grandson is lounging about doing nothing and he's soon on eBay looking for a mower. Since he found me what has turned out so far to be a good car, he's a bit full of himself. He's soon found a mower and we've bid £500 for it, but I'm a bit uneasy about it.

I sleep on it and next day I ask him to pull the bid back out. 'Can't do that.' 'What happens if you make the top bid but you don't follow it up and buy it?' 'The police come and get you.' Luckily someone tops our bid by £10 and I breathe a sigh of relief. He might very well be good at buying cars but he could be crap at lawnmowers.

## MAY 5TH 2012

I've got this second-hand lawnmower on trial and I quite like it. It's bigger than the one that died, it's in better nick and it's doing a very nice job. I ask a few questions about it, it's about the same price as a new engine for my old one and I am quite tempted. So I ask a few more questions about it: what's its age, where does it come from, the sort of stuff Cilla Black used to ask on her dating programme. They are a bit vague about its age and even vaguer about where it comes from – there's a wave of the hand: 'Someone the other side of the hill.' There's lots of hills around here so this is no help at all but I don't really need that sort of help because I already know that it came from a nephew of mine who wanted something a bit smaller for his young son to drive. What the dealer doesn't want is for the two of us to get our heads together and work out the deal he has done and the deal he is trying to do with me. But he's too late for that.

Of more immediate concern is its size. It's all very well having a bigger machine and getting the job done quicker but it has a two-cylinder engine and I wonder how much petrol it uses? And who would ever have thought that there would be a time when we would have worried about that? Have you seen those old photographs of horses with sacks on their feet pulling lawnmowers? It's only a matter of time.

\*\*\*

When my son was single he would disappear with his rugby kit late on Saturday mornings and reappear sometime on Sunday mornings. The time that he reappeared would depend on all sorts of things, such as where he'd played rugby, where he'd crashed out for the night. Because licensed premises would have been involved, that would have had some bearing on what time he considered it safe to drive home. We were always interested in what had gone on whilst he was away, starting with the rugby match itself and then what had gone on during the evening out. We called it 'giving his report'.

One Sunday morning he 'reported' that as he had returned home that morning he had seen a camel's head sticking out over the hedgerow looking at him. We teased him that he'd had more to drink the night before than was good for him and that he'd imagined the camel. But we knew he hadn't – there's been a small family circus around here for as long as I can remember and their animals have spent the winter at various locations in the neighbourhood ever since. They were performing just up the road last week and I took two younger grandchildren to see them.

A circus with performing animals is controversial with some people these days – what isn't if it involves animals? – but for children who spend so much time with toys that are linked to modern technology, to see their delight in such simple, basic fun, makes it all worthwhile. And if the ringmaster in his top hat and splendid red coat looks a bit like a person in the kiosk who took your money as you came in, and bears a close resemblance to a little boy who went to the village school 30 years ago, it's because it *is* the same person. Just the same as the lady on the tightrope looks very like the lady who sold you your popcorn, who looks like the little girl who came to your daughter's birthday parties years ago and it's quite confusing because she looks exactly like the lady going around the ring bareback on a horse.

So I settle down to watch. I watch the circus and I watch the children's faces. The youngest boy I have with me is quite nervous to start with because we have ringside chairs and are very close to the animals. But he's soon bold enough to go into the ring to meet a Shetland pony that can tell children's ages by pawing the ground. The pony thinks he's six and he's actually five but his elder sister tells the ringmaster that the pony has done very well because he's six in a fortnight anyway.

Having a local circus has always been 'interesting'. There's the incident with the camel's head through the hedge and over the years, there have been escapees. Local people have always taken it in their stride, very much as if someone's sheep are out. 'Newcomers' have always taken meeting baboons on the road more seriously and ramblers who come across a camel and a couple of llamas are on the phone to the police in a trice. The circus used to have a brown bear and you would often see him being driven about sitting in the front seat of a Land Rover, belt on of course.

But there was a third dimension to my visit to the circus. There was the obvious pride in what they do of successive generations of the family that run it. It can't be easy these days keeping all those animals looked after and all that kit to move about.

Mindful of the many critics that circuses have, I looked very hard at the animals. After all, I do know a bit about them. Firstly their condition, regardless of species. They were all in fine fettle, not too fat and nowhere near thin. Then I looked at their feet: feet are important. Their feet, again regardless of species, looked perfect. Then I looked for signs of stress and could detect none. We had a wide spectrum of species, from pigeons, cats (though not at the same time), dogs, goats, ponies, horses, a zebra, snakes, llamas and a camel (probably the same one) and nowhere could I detect a stressed animal. The only animal that looked unhappy was the fox riding the mule around the ring, but if you were a fox and you knew the hunt kennels were only two fields away, would you look happy? So I know the animals are well and happy, I know the family who own them to be kindly and friendly, so where's the harm? Because it would appear to be harmless, yet these circuses do have their critics.

There are those that would ban them. There are plenty of people about that would ban things. And if you want to ban things with animals involved, a circus is a relatively easy shot. It's picking at the low fruit before you move on to your next objective, and there will always be a next objective. Who is to say that getting animals to perform in this way is any worse than tipping a cow on her side to trim her feet, or running a horse at Cheltenham? It's probably not, but that doesn't mean they should all be banned. Someday, some time, those of us who see ourselves as fairly normal, will have to say that enough is enough and make some sort of stand and stop this headlong journey along the animal welfare vegan road.

## May 12th 2012

It was Black Friday for Dairy Farmers last week with the news that a major milk buyer is about to cut its milk price by two pence per litre. The rest of the buyers will follow suit so quickly you won't believe it. This could be the defining moment for some dairy farmers, the final straw. I don't expect any public sympathy because a lot of folk are having a tough time at present but the difference is the animals. If I made nuts and bolts I could switch the factory to a three day week or close it down, knowing that in better times I could start it all back up again. But now lots of good milking cows could end up as beef burgers for the Olympics over the next weeks which is a shame and a waste of valuable resources. If you took a long-term view of life, you wouldn't let the national dairy herd reduce in this way, but people take short-term views and dairy cows and dairy herds will be fewer, never mind the dairy farmers.

When I was first bitten by the farming bug you could probably make a living on a rented farm of 50 acres by milking 20 cows and keeping 200 hens in a deep litter shed. You might keep a few sows instead of the hens or you might keep a few sows as well, but there was a living to be had. This is very much how I started farming albeit with 80 acres, 30 cows, 200 hens and a few sows. My mother-in-law bought me the 200 hens as a wedding present but there was no money in eggs at the time so I sold them back to her. But pigs were OK. You could sell eight weaner pigs off a gilt and they would make a fiver a piece. Sell two litters of pigs and you had enough money to buy another cow. It wouldn't be the very best cow, but it was a cow nonetheless.

Within about four years I was milking 100 cows on my 80 acres and buying most of their winter diet in the form of straw, a diet, I might add, that you wouldn't dream of feeding today. At the height of my output I was selling 240 gallons of milk a day out of the milk tank and putting ten churns of milk on the communal milk stand in the village. Three others used that stand and my ten churns were more than theirs altogether. This was the stuff of legends, this sort of output off a small farm.

What the story serves to illustrate is that this is how it has always been, requiring more and more output, just to stay in the same place. The 80 acres is still my base unit and it's not been enough for some time. So you have to try to enlarge it and the only way is to rent more ground. That's what I've done over the years.

I've had bits of land all over the place during that time which is what this particular story is all about. But we're not ready to get to the purpose of the story just yet. It's a great feeling when your son decides to join you on the farm. You always say you didn't push him to do it, but you probably, very subtly, did. So that's great, he goes to college, and now he's back home and he gives new resource to the business. And he's happy as well. All he wants in life is a Ford Escort, plenty of beer money and his mother's cooking. But gradually that changes. It usually changes when the regular girlfriend turns up. A few years later and he's got a house and children and the farm is stretched to cope with keeping two families instead of one. It was the search to make the business bigger that took us into poultry some years ago, establishing a unit on the farm that didn't use precious acres. Except that 200 hens didn't do it anymore, we had to keep tens of thousands. Now we can get started on the story.

## MAY 19TH 2012

I've just come back from a week's family holiday. I organised it and it took some organising. My sister-in-law wanted to go on a cruise. My sister didn't want to go on a cruise because she unfortunately is in a wheelchair and the main purpose of our holiday is to get her away whilst she still can. She didn't want to go on a cruise because she wasn't sure about handling a wheelchair on a ship. My brother didn't want to go with my brother-in-law and my brother-in-law didn't want to go with my brother and my wife didn't want to go at all. So I thought I'd done well to organise all that.

We went to the Italian lakes again. After two days I could have cheerfully hit my brother-in-law, and after four days, my brother, so when a boat man on a ferry fell out with me on the fifth day, there was no way he could have known how close he was to ending up in the lake. If ever you get the chance to go to these lakes, take it, the scenery is spectacular. But it's the small detail I tend to focus on. In a week I only saw three animals in a field and they were horses. I saw the back leg of a goat but that was cooking in a fast food outlet in a street market. I've never seen such a clean hotel. It was pristine, clean sheets every day, if a petal fell off a flower in reception, the receptionist would come rushing out to pick it up. It was a small modest hotel, reasonably priced, immaculate in all respects except one.

There was a very small bar in the lounge and on the wall at the corner, the corner that sticks out to join the wood of the bar, was an endless two-way procession of ants. It fascinated me. There were always a hundred in view, some going up, some coming down, I couldn't work out which way they were carrying and what it was they were carrying. They went under the wood of the bar and they disappeared into the ceiling, and as far as I could tell, no one paid them a scrap of attention.

\*\*\*

We've got this free-range cockerel, his name is Neville. He attacks people. He is quite good at attacking, he sneaks up on you and he has been known to knock over unsuspecting grown men, including me, as they carry a couple of buckets of calf milk. Grown men don't like rolling about in the mud with milk all over them, it's not very dignified and it's not very good for your ego when the story is related in the pub at some later date, although everyone else thinks it's very funny. It's all worthwhile in a way, because he attacks the unsuspecting visitor, he's had several salesmen and who knows he might get a farm assurance man one day.

But I worry about Neville's future. I'm worried about him attacking a small child and I'm worried about the possibility of retribution. He has a clearly defined territory and I have put an eight foot piece of hazel stick conveniently against the gate post. It has a whippy end so I can drive him off without doing him any real harm, except to his ego. There are people here who have threatened him with heavier objects and I am concerned about his possible demise.

So I try to work out a strategy that will save Neville's life. Years ago we had a free-range turkey stag called Boris. Like

Neville he became one of the family. He used to come to the kitchen door every morning for some toast and every Christmas day I would stand him on the kitchen table for a minute or so, then get him off again. I would tell him that not many turkeys could say they had done that. He danced the night away at my son's wedding in a marquee on our lawn. He ended up inside a fox but I still miss him. So my thought process is that, if I get another turkey stag, he will distract Neville from his human attacks and inadvertently save Neville's life. My daughter works at the farm museum down the road where they filmed Victorian Farm. They've got turkeys (and it's Father's Day soon).

A few days later and I'm on my way to fetch some turkeys. We can't find any boxes to put them in so I just put them loose in the stock trailer. I get a stag and three turkey hens, big black ones. I leave them in the trailer for a couple of days to settle in, then I put them behind a gate in the calf shed. The hens are very tame, you can pick them up to stroke, and they can get out straight away but the stag is not as agile so it works quite well. The hens explore their new home but they don't go too far because the stag calls them back. He gets out next day and the day after that they are all on the main road but they've settled down now and stay on the yard.

And so far the plan works. Neville has cut down the size of his territory and he is completely distracted by his new wives. But there's a bonus, the turkey hens are laying, and by Saturday night I am taking orders in the pub for Christmas dinners. There's great merriment in the mixed company as they hear the story of Neville and the turkeys – they all know about Neville in the pub; the landlord has written a poem about him. Then we move on to more practical issues. Where will I hatch out the eggs? 'Have you got an incubator?' 'No.' 'Have you got a broody hen?' 'No.' In the end I suggest that there would be no better place to hatch some turkey eggs than within the bosom of the young lady sitting opposite me. She, and all there, agree that there could not possibly be a better place, she assures us the eggs would be warm and comfortable. So that's sorted, I'm just making a list of volunteers to go and turn the eggs every 24 hours, there's quite a lot of volunteers. This isn't meant to be sexist, boobs are boobs and turkey eggs are turkey eggs, a fact of life. And you've got to hatch them out somewhere.

## May 26th 2012

I've spent most of the last week on a tractor working ground down with what we call a power harrow for crops of turnips and kale. It's a slow dusty job, I only travel at about 2mph and it could be boring if it weren't for the wildlife. There's eight hares on the field and I wonder why they are spending time on a bare ploughed field when all about them is greenery. It could be they have leverets there, but I've not spotted them yet. Then there are the rooks, dozens of them, looking for grubs that I disturb as I pass by. Then a buzzard turns up. At close quarters, and you get up close on a tractor, they are quite an attractive bird. With each pass of the tractor another turns up and we are soon up to eight, if it keeps on like this it will get scary and I'm glad it's tea time and time to go home. The final job is to roll it all down. The rooks all turn up because they have seen the tractor but with the roller on I'm not disturbing soil and they can't work it out: why no fresh grubs? So they stand and watch me go up and down and I can tell they feel cheated.

***

It's after milking on a Saturday afternoon and two of us are taking 25 turkeys eggs to put in an incubator. It's a pleasant task. It doesn't need two of us to transport 25 eggs but we are going to the local farm museum where the turkeys came from and that's always worth a visit. A couple of 100 yards from a neighbour's farm, this big white van pops out in front of us, and speeds off. 'Wonder where he's going, looks as if he's late.'

This farmer does pig roasts and although we keep about 200 yards behind him, our truck is soon full of the smell of roast pork. So much so that Mert stirs himself in the back of the truck thinking he's about to get a treat. The 200 yards distance we keep behind him is very critical. Two weeks ago another farmer on a similar errand went over a bump in the road a bit quick and the back door of the van burst open and a roast pig popped out onto the road. But there were no rich pickings to be had because there was an articulated lorry very close behind and it quickly flattened the pig out for the magpies and crows. We are not about to make the same mistake and if this pig drops out in front of us we at least expect to get some crackling. But it's not to be our day and we have to turn off. The smell lingers in the truck for a long time. Wonder if you can buy roast pork air-fresheners?

*** 

We've got this gateway into a field. It's off a narrow unclassified lane, the grass verge is wide there and from the evidence, it's a popular place to stop. It's a very quiet road and the view from there is wonderful. If you are not used to the view, it is probably spectacular. The evidence that it is used a lot is disappointing. The evidence is litter. It accumulates on a daily basis, fast food wrapping, bottles, you name it. I pick it all up, I don't think I should have to but I dislike picking it up less than I dislike seeing it. So last week I'm going past there and there are a couple having a picnic, and a very dignified picnic it is too. Sensible Volvo car, two folding chairs at the rear, a nice little folding table, quite a pleasant scene.

When I come back by half an hour later they are just packing up to go. The chairs and table are going into the back of the car but all the debris off the picnic table is still on the floor. So I stop. They are a couple in their late 60s, (nothing wrong with that) quite poshly spoken, (probably from Bath) and I point out the litter they have left and ask if they are going to leave it there. They seem very surprised that I should even ask the question, 'We don't want all that rubbish in our car.'

This is a testing time for me. I tell them that if they don't pick it up I will phone the police and that we, the dog and I, will make a citizen's arrest until the police get here. They pick it up with very bad grace; they can see that I'm very angry, though I've not lost my temper. I notice that the rubbish is put on the floor on the front passenger side so I bet it's back out of the window as soon as I'm out of sight. I'd always assumed it was young people doing this; quite surprised that it is all ages.

Next day, same gateway, and there are three Tesco bags just inside the gate against the hedge, they are out of sight of the road but it is litter nonetheless. I get out to move it with a sort of groan. But the bags are quite heavy so I look inside. They are each full of pre-packed cheeses, quite a lot of cheese in fact, the three bags probably weigh 25lbs. It's curious, but my mind soon works it all out. It's obvious to my fertile imagination that this cheese has been dropped off by a home delivery van driver, who has nicked it at the place he works and has left it here to be collected by an accomplice. I put it all back exactly as it was and decide that I will keep a closer eye on this gateway for the next few days, I think they call this sort of thing a stakeout.

A week later and the cheese is still there so I pick up the bags again for a closer look. It's all four to five years old! I have to have a solution, so I decide that someone has been clearing out a freezer. Quite why they have to drive out into the country to dump it is beyond me. By now the mice have made a start on it, so I leave it there for them to finish, it's still litter but I expect there will be some more tomorrow anyway. Some of the litter, I can't even tell you about.

### JUNE 9TH 2012

Though we've plenty of hares on our away ground, they are not so plentiful here at home. I saw three at home two years ago but not one last year. Stephen was out putting fertiliser on our grazing land this week and I was delighted when he reported seeing four adults and several leverets. There were a set of tiny twins, 'no bigger than little kittens' he said. As he moved across the field the little leverets would scamper a bit further on as they sought fresh cover in the next clump of grass. One was only ten yards from his tractor wheel when it was snatched by a buzzard and carried off to its nest in the adjoining wood. I've seen this happen several times. I always find it upsetting.

***

As a part of the distraction needed to steer pub conversation away from foxes and the safety of important cockerels, we are now full on with shearing stories. Groups of men working together invariably produce anecdotes and an element of fun but there's no glossing over the fact that it is desperately hard work, mostly done, these days, by gangs that travel around shearing huge quantities of sheep. It's hot, hard, dirty work and I wonder sometimes if some of these top shearers will suffer in later life with all that bending and exertions. There are plenty of young shearers who go to Australia and New Zealand to shear as well, making it a 12 month occupation.

I always remember a first rugby match of the season and we were playing a team from a city in the Midlands. One of our centres was a bit late, he'd been shearing that morning, a slightly built lad, wiry but very strong. He lined up in his position, he looked a bit tired, there was still sheep muck on his forearms. His opposite number was almost as big, strong, had a nice suntan from his holiday abroad, bit of a poser, he'd got the collar of his shirt turned up (always a giveaway), I don't think he could believe his luck when he weighed up his opponent. First chance he had he put the ball under his arm and ran straight into our man. Our man didn't really move; his opposite number ended up on the floor whimpering and had to have attention before he could carry on. Another player asked me what had happened. I said, 'Well from where he comes from I don't expect he meets many people who can shear 400 ewes a day.'

I didn't start to keep sheep in any sort of numbers until later in my life, but I was determined to be able to shear. I used to get two others to help and we sheared in a field with a machine driven by a petrol engine. None of us were brilliant shearers so we sheared two heads and would catch our own ewe, shear it, wrap the wool and then it would be your turn to catch another ewe. Unfortunately the field was next to the pub so I would run a tab for drinks and crisps at lunchtime, drinks after we finished, same again next day. With the wages, the food and the drink, it was an expensive two days' hard work. My helpers couldn't come one day, and a very shy, quietly spoken Welshman in his mid-forties slipped unobtrusively into our yard. He'd done the lot by 2.30 in the afternoon. His bill was just about half of what it had cost me the year before; it had taken less than half the time and all I'd done was get the sheep in, wrap the wool and put a red 'E' on their backsides.

It's hard work catching sheep, far better if you can get someone else to catch them, and I remember the story of a friend of mine, in fact it was he, the rugby player. When you shear, the last thing you want is the sheep straining against you all the time, so an important part of the skill is how you balance the ewe and move her about while you remove the wool. One of the worst things that can happen while you shear is for the ewe to get away before you have finished. She does a runner, possibly out into the field, half her wool is dragging along behind her and half of it still attached. There's usually a mad scramble to catch her again before there's bits of wool everywhere.

My friend was shearing with a gang of three others at a place he described as 'up in the hills owned by two little old boys'. These 'two little old boys' couldn't resist the golden opportunity presented by someone else catching their sheep so as the ewe sat on her backside being shorn, they would attempt to trim her feet at the same time. So as the shearer tried to control his sheep, every few seconds she would struggle and kick because her feet were being trimmed by a penknife. It wasn't long before the shearers had had enough of this and at a given signal the four of them let their sheep go. The little old boys couldn't catch four ewes at the same time; they ended up with four half-shorn ewes out in the field and handfuls of wool everywhere. They had to get their dogs and get all the shorn sheep back in, in order to catch the four while the shearers sat down and had a cup of tea. There was no more foot trimming done after that. I don't miss shearing, life has moved on, mine in particular, but I do miss sheep washing. Nearly everyone used to wash their sheep a week or so before shearing, years ago. The wool would 'rise' and they would shear easier. There was a pool or a place where you could block a stream in most localities and usually someone would end up in the water with the sheep. I don't know of anyone who washes sheep now. Think I'll suggest it in the pub. I bet 'they' wouldn't let you do it these days.

### JUNE 16TH 2012

The fox has had two of my turkeys so I've only got the stag and one hen left. 'Serves you right,' they all say, but I wanted my turkeys to be truly free range and they settled down nicely in the calf shed roosting at night on a gate. Now I have to go up the yard every night and shut them in the stock trailer, which wasn't the original idea. And it isn't 'them' any more either, it's just the stag because the remaining hen has disappeared and we presume she has some eggs somewhere and she only turns up once a day for some food. We've looked for her everywhere but just can't find her nest. You can't get much more vulnerable to fox than sitting on some eggs all night, goodness knows where. The fox, or foxes, are only doing what foxes do, I just wish they wouldn't do it around here. And there's plenty of rabbits about, why don't they get them?

***

We've changed my wife's car. We moved her up about six years and moved her down about 80,000 miles for a modest three figure sum. 'This new car is very good on petrol,' she says. Next day she runs out of petrol on a major trunk road. She'd been looking at the temperature gauge.

***

So the Queen's Jubilee has come to an end, gone. For reasons I'm not sure of, I was on the village committee. I was put in charge of risk assessment, which everyone thought was hilarious, so hilarious that I started to find it quite hurtful. The Monday evening celebrations, pre beacon lighting, were a disco and hog roast at the pub. I was put in charge of events at the pub as well. 'Roger is always in the pub, he can see to that.' Then something happened that makes it all worthwhile.

A 'newcomer' comes up. I've never met him before but decided some time ago that I didn't like him. 'Are you Roger Evans?' 'Yes I am.' 'Do you live at _____ Farm?' 'Yes I do..' 'Well I've been to your farm twice now with poppies for Remembrance Day. I've never got close enough to deliver a poppy but your dog has bitten me twice'. I tell him that my dog is a good judge of character and this produces a scowl. Makes me feel good knowing that Mert continues to live up to his reputation. When I get home Mert gets a pork pie for his supper. Anyway I get my poppy off a girl in the village, he should know that.

### JUNE 23RD 2012

I told you I had rented some 'new' fields and a large cattle shed, didn't I? Well the farm where it is, has been split up, and there's another man rents the farmhouse. The man in the farmhouse has a few hens roaming about the yard and one of them started to lay in the shed of one of my neighbours. He was delighted with this and went out and bought a dummy egg to put it in this hen's nest. So the hen goes to her nest everyday and there's always an egg there (even if it's a rubber one) so she keeps on laying there and as hens aren't big on counting, she doesn't realise that the eggs don't increase in number.

We get the story every Saturday night in the pub, it's told with great glee because the husband and wife are getting a free egg every day. 'We had six lovely eggs again this week, beautiful they are, lovely brown shells.' 'You should see the yolks, really rich, golden they are.' And husband and wife go 'mmm' to try and convey to us how tasty the eggs are. I had a look in their shed one day and I couldn't find the nest. But you just have to be patient in life and the story will eventually unfold. On Saturday they say 'We had two eggs this week, she's laying between two bales of straw and I've got some black plastic sheet over them and the hen pops under the sheet to lay but no one can see her nest.' I find the nest easily next day, there are two real eggs and the rubber one, I remove all three and replace them with two Cadbury Creme eggs.

There's been no mention of eggs since but if you watch their faces, you can see their eyes darting about looking to see if someone will give the game away. I look in the nest again four days later, the hen is still laying but the chocolate eggs are gone. I put the rubber egg back in the nest and in a minute I'm having the two real ones I lifted for my breakfast. I haven't decided yet what the next step will be, I'm trying to locate a big left-over Easter egg. Or hard boiled eggs in the nest could be the next answer.

*** 

I don't know how we got around to the subject, but get around to it we did, we were discussing National Insurance contributions in the pub. Thursday night in the pub is farmer's night (and Saturday, Sunday and Tuesday nights) but Thursday night is the main night. It's only really dedicated farmers like me that go four nights. Sometimes non-farmers pull up a stool and try to join us but I always tell them they will struggle intellectually with the conversation. We talk about tractors and sheep and the price of red diesel, heady stuff like that. Now I think about it, I think we came to National Insurance via banking on-line and direct debits and stuff like that, then someone says, years ago they used to have to buy stamps and put them on a card.

Then they all look expectantly at me for further explanation. And it's true, we did have a card and there was a place on it for a stamp for every week of the year. You purchased these stamps at the Post Office and they were quite expensive and you had to keep the card stamped up-to-date. And of course you didn't. Then these people would come around on random checks to see your card and if your card was blank you were owing quite a lot of money. So they would make you stamp it up-to-date 52 times, a lot of money, might be more money than you actually had in the bank, so not only would you get a bollocking off the person who came to check the card, you would get a bollocking off the bank manager as well. In the end a lady would call here every three or four months to see my card, she knew it wouldn't be a wasted journey, and she would tell me off and I would have to get these stamps and she would call the next week to see that I had.

She called one day and I was five months behind and she must have been having a bad day because she was quite nasty about it. 'You know you've got to buy these stamps: why don't you do it regularly, I think this has been going on too long, I think I'll report you. Why is it so difficult?' Looking suitably contrite and with a flash of inspiration, I tell her I can't read or write. She is appalled, not with me, but with herself for being so cruel and giving me such a hard time. She can't apologise enough. So for the next four years, until she retires, she calls here every week, collects a cheque, takes it to the Post Office in the village, gets a stamp, comes back here and puts it on my card. We become quite good friends and when she retires she is clearly worried about how I will cope in the future. But the future soon brings in standing orders and direct debits which take over from stamps bought at the Post Office, which is where all this started. I don't tell them all this in the pub, I wouldn't want anyone else to know, I still feel a bit guilty about it.

*** 

I expect you all know the story about the dairy farmer who won millions on the lottery. He was asked what he was going to do with the money. He said he would just carry on milking cows until it was all gone. I was looking back at some stuff I had written 30 years ago. I'd said that the only way you could end up with a million pounds from milking cows was if you had £2 million to start with. Obviously written in pre-lottery days, but quite prophetic. The sad thing is that most of us keep on milking cows because we think that it can't go on like this forever, it will have to get better. And here I am, 30 years on, still thinking the same thing.

## JUNE 30TH 2012

It's my first walk up the yard on a Monday morning. The first thing I notice is that there's a lot of ravens about the yard. That's an important distinction, there's always ravens about, but these are about the yard. They are busy, very animated. They are not sitting on wires like swallows wondering where Africa is. They are busy behind the calf shed. Stephen is feeding the calves but they are not bothered. There's a turkey got a nest with nine eggs in it, just behind the shed. They've driven her off the nest and eaten the eggs. A bewildered turkey appears around the corner and the ravens disappear to wreak their carnage somewhere else.

***

The conversation in the pub drifts on to sheep getting out, there are usually persistent offenders and they will usually get over a fence or make a hole in the hedge and all the others will follow them out. I don't know if you'd be allowed to do it now, but we used to put devices on trouble-makers to stop them breaking out. This was a beef farm before I came here and they kept beef cows that suckled their own calf.

If a cow had too much milk for one calf they would get it to adopt another calf as well. I found, in a shed, some leather collars joined by chains, a sort of handcuff effect, the idea was to put a collar on the calf and the other collar on the calf to be adopted so that the two calves became inseparable and the adoption would proceed quite quickly. Once the cow had accepted the calf, the collars could be removed. They were well made with a swivel in the middle of the short chain so the chain couldn't twist and strangle the calf. I cut these chains in half and stapled the loose end to a piece of stick about a metre long, then I would fix this device around the neck of a ewe that was always breaking out. The ewe would very soon get used to this stick around her neck and also very soon get used to the idea that it was impossible to get a yard of stick through a foot wide hole in the hedge. I never thought of it as cruel but I suppose it's all a matter of perception. Some people would cut a forked piece out of a branch and fix it around the ewes neck with a third piece of hazel stick so that she had a sort of three cornered collar. I haven't seen one of those devices for years and years.

## July 7th 2012

It's early Monday evening. The First Responder turns up at a house in the village. Next it's the paramedics and then, to great excitement, the Air Ambulance drops in, the small field belonging to the school right in the middle of the village. It's something of a false alarm because the patient is able to walk to the ambulance that comes next and is happily discharged from hospital the next day. On Wednesday evening we are at the monthly diners' club at the pub and there is much mumbling and grumbling about the cost of sending the helicopter out and 'they wouldn't be able to do it if people didn't raise so much money for them.' That's a bit of a shame really because it's almost instinctive for people to have parties and anniversaries around here, where they donate the proceeds to the Air Ambulance. And why not? It's made such a huge difference in remote rural areas. Countless lives have been saved with the combination of mobile phones and helicopter in cases of accidents in fields or woods or sudden illness in the countryside. It would be a shame if the appearance of the air ambulance at less urgent cases should blight future fundraising. I hope it won't.

Once the helicopter was stood down, the crew showed the assembled children around it. When it left it did so with something of an aerial flourish. If your garden was on the periphery of the downdraught of this flourish, it dried your washing quite nicely. If you were nearer to the epicentre of the flourish your washing came off the line and was scattered to the wind. It apparently took a couple of days to get all the washing back to its rightful owners.

One of the larger ladies at the dinner tells us that there is still no sign of her Sunday chapel knickers. 'I reckon I know who's got them,' she says, 'and it isn't a woman.' And everyone nods knowingly as if they know as well. This is a bit annoying as I haven't got a clue who they are talking about.

***

As I get older, or should I say old, I seem to be getting more patient. I waited for two weeks for someone to reveal where the nest was where someone else's hen was laying them a daily egg. I removed the egg one day and replaced it with two Cadbury Creme eggs. A week later I put in a hard boiled egg. For another two weeks, nothing. Then at the same dinner at the pub, heads are lowered and the story of the chocolate eggs comes out. The story causes great merriment: most of the company know it's me, but the story isn't finished. 'Next week he brings home an egg with a bloody lion already stamped on it.' More laughter. She's obviously not tried to crack it into the frying pan yet. I've got two minute bantam eggs to put there next week. This is a story that will run and run until someone gives me away.

### JULY 14TH 2012

We've all been told of further price cuts that will kick in on 1st August. The majority of dairy farmers are in despair. Sainsbury's pay their dedicated suppliers just over 30 pence per litre, a figure based on the cost of production, but there's only 2-300 producers lucky enough to get this. Most of us are on 24p a litre which is six pence less than Sainsbury's reckon it actually costs to produce. In real terms that's over £70,000 a year on this farm, money which could be spent on giving men and animals a better life.

We needed some brake parts for our number 1 tractor. We made six phone calls to find the cheapest place. Stephen fetched them, he put them in one hand and they cost more than a day's total milk output. I just feel so angry. I once believed there would always be a living producing nature's most natural and complete food and because of that I've milked cows for nearly 50 years, constantly hoping things would get better. I never thought that I would think this but if every other farmer would dump their milk for a few days, I would dump mine. I'm angry enough to tip it in a river, never mind spread it on the land, and if the Severn went white and you could see it on a satellite photo, so what?

### July 21st 2012

Yesterday I went to London. I went on a bus with other dairy farmers. We went to protest about recent price cuts on the milk we sell and those further cuts proposed for 1st August. I've never been on a protest before of any sort and was amazed at how many people turned up. The attendance was put at 2,500-3,000 and we were, I suppose, showing an immense solidarity. I went because I was angry.

I was still angry when I got home. Our budgets for this year showed a very tight year ahead with the price of cereals and proteins at record levels. If we apply price cuts to those budgets, we have to borrow a lot more money to get through the winter. How would you feel if you had six months' hard work in front of you and the only prospect at the end of that six months was that you were tens of thousands of pounds worse off? It was all peaceful enough but at the back of my mind was the thought that we might have done more good if we had all gone and sat in the road outside parliament.

And that sums it all up: that someone like me actually contemplates very seriously taking direct action because of the way I and other dairy farmers find ourselves being treated. Next time you buy some fair trade coffee, ask them if they have any fair trade milk? They will tell you they do, they will tell you that they have a dedicated milk pool and that they pay their dedicated dairy farmers a price based on cost to produce. And only about 12% of the milk and dairy products comes in to that category.

***

Today's is a delicate subject and I will deal with it as sensitively as I can. Today we take a peep under ladies' skirts. But before we get to that I have to set the scene. When I started out as a dairy farmer, the dairy industry seemed to be dominated by ladies. These ladies seemed to be of an age that was described as middle onwards. They were fearsome ladies and all the ones that I knew were of single status. The lady in my story was the one who decided where, how, and very importantly, if, you could milk cows. She was the guardian of hygiene and food safety standards. Her word was law.

My milk went to one of two dairies where it was put into bottles that mostly ended up on doorsteps and both of these dairies were run by similar ladies. The tanker driver spoke in awe of them. All this, to put some perspective on it all, was 48 years ago, which was when I started out as a dairy farmer and was in my early 20s. The timing was important, because it was pre miniskirts and so we come to the delicate bit.

With miniskirts came tights. My memory is not what it was but I think I can fairly safely say that these ladies in the dairy industry never wore miniskirts, and they would have worn stockings. How do I know that? We'll find out very soon. Stockings left a gap of bare leg between the top of the stocking and the next under garment, so it became fashionable, at the time, to wear knickers with some leg on them that covered this gap. These knickers were often brightly coloured with a pretty lace edge around the bottom of each leg. (I'm beginning to wish I hadn't started out on all this). So, I've set the scene, I've been lucky enough to find a farm and I have to set it up to milk cows. I have to put in a parlour and a dairy and that is the moment that brings in the dairy advisor who is employed by the government.

She wants to know everything, where and how I will milk, the construction of the dairy. What is the quality of water I have, where do the drains all go? I find that I get on really well with her, she's very helpful, my young wife reckons she fancies me. So one day she wants to see where the drains from the yard go and where the septic tank is and as a result we find ourselves at the bottom of one field and have to go into the next. I suggest that we go back to the gate but she will have none of it and says she can easily get over the barbed wire.

And so finally we bring long tweed skirts and long legged knickers together. She hitches the long skirts up above her knee and puts her leg over the barbed wire. The skirt may be long, the knickers may be long but the legs are too short. A leg of a knicker becomes attached to the barbed wire. There's no panic about this and at this stage she gets very giggly. We can't see the intricacies of the problem because it is all under the tweed skirts. But she's a very practical person and works out that either the knicker has to come off the barbed wire or the knicker has to come off the lady. She can't do either of these tasks herself because she's balanced precariously on her toes with both her hands on my shoulders.

So this young sensitive boy has to delve under the skirt and unhook garment from barbed wire and then help to keep garment away from said barbed wire whilst the other leg goes over the fence. My lady advisor enjoyed the experience immensely and I thought that her hand stayed on my shoulders just a bit longer than was strictly necessary. This experience had a profound effect on my life and I've had to have counselling for several years. I've been afraid of women ever since. Wonder what made me think of it now?

**JULY 28TH 2012**

So what of the turkeys? Well we've still got the stag. I deliberated for ages about a name for him and one day, without thinking, I called him Eric. It seems to suit him so that's one job sorted. We only have one hen turkey left and she disappeared for a couple of days to sit on some eggs but we found the nest before the ravens did and moved them to safety in an old stock trailer. She stays in there all the time and sits on her eggs. We put Eric in there at night to keep him safe from the foxes. And all is not lost in our turkey world because safely in a nice coop with an outside run, we also have seven young turkeys. They must be a month old now and Eric spends much of his time parading up and down, showing off to them puffing out his feathers. I suspect that he has carnal intentions towards these turkeys which isn't good because they are his children, but if you are a turkey you don't necessarily know that. Some of these young turkeys, at least three of them, will become Christmas dinners and they don't know that either. But there's only so much posing that Eric can do to baby turkeys and with his only remaining wife busy sitting her eggs, he makes his way down to the house where there are good reflections to be had in cars and kitchen windows. And that is the only real company he has at the moment: a reflection of himself in our kitchen window.

*** 

It's ten o'clock at night and I'm in the pub. There's some people camping in the field at the back of it and a couple of them come in now. One is carrying a dead rabbit. 'I found this dead on the road,' he says. We all see dead rabbits on the road at this time of year, it's no big deal, the ravens, crows and magpies hoover them all up every morning.

This rabbit is getting some attention because there's not a lot of road kill makes it into the pub. He's not dangling the rabbit by his legs, he's cradling it in his arms as if it were the baby Jesus and he was just about to put it in the manger. I think to myself, 'Oh no, he wants a rabbit funeral.' But he doesn't, 'Would one of you skin it for me please, I want to eat it.' 'Put it down here,' someone says 'I'll skin it for you in the morning and bring it back down tomorrow.'

And he does just that, does a tidy job of it, and takes it back down at lunchtime, in a nice clean polythene bag. But he's not thanked very much, in fact the camper is quite indignant, 'Where's the skin?' 'It's in my wheely bin with the rest of it.' 'I wanted the skin!' 'What do you want the skin for?' 'I wanted to make it into a hat.' So my friend has to walk back home, delve into the wheely bin and bring back the skin. And the camper takes it off him triumphantly and turns to walk to his tent. Nothing wasted in that family then.

\*\*\*

There was a beer festival in our local town recently. There are lots of beer festivals on lots of local towns, all over the place. They are an interesting phenomena. First up we get the beer connoisseurs: you mostly need to have a beard to be one of them, you will probably wear jeans or corduroys and you will be boring, because all you can talk about is the beer. You will probably go to a beer festival every weekend in the summer. You can't believe your luck that your wife doesn't seem to mind, I bet she doesn't mind: while you're away every weekend she probably is having an affair with your next-door neighbour because he is infinitely less boring than you are.

Anyway, mid-morning you, the beer bore, turn up at the beer festival and there is a vast array of beers available mostly with unspeakable names or describing doing something illegal to a sheep. And you stand at the bar and wait to be served and when it is your turn you haven't made your mind up what beer to try so you stand there licking your lips, 'Now what do you recommend, landlord?' and the landlord stands there while you make up your mind and gives you one of his thin smiles which means he's had enough of the beer festival already and it's still only 11.30 on the Saturday morning.

Other customers are clamouring for attention but you deliberate a bit longer before you decide, 'I'll try half of that Gypsy's Underwear please.' And you get served, but you delay things a bit longer yet. You're also one of those men who keep their change in a purse (no wonder your wife is having an affair) so it takes a couple of minutes to count out the change and by now the landlord could quite cheerfully kill you. My old Dad told me never to trust a man who kept his change in a purse or did the three buttons up on his jacket and he was right.

As the day goes on, the festival goers start to change. Mini buses from other local towns arrive, carrying serious drinkers who don't dither about deciding what they will drink, they will have a pint from that pump on the end and they will work down the line of pumps from there. And gradually they will become the majority of drinkers and your half pint drinker will have gone back to his tent for a bit of a nap that will last until next morning. At about 7.30pm the fights will start. They will be serious fights between men whose only difference is that they live here and he lives ten miles away. And because they are big men and have had a lot to drink, windows are smashed, furniture is broken, people have limbs broken and the mini buses that are arriving now carry police reinforcements.

I hear all this second hand in the pub on Sunday night: by now there have been three ambulances and the police were riot police, I don't know if it's true, but it makes for a better story. What is true is that I can easily see all the broken glass on Monday morning. Nothing like a good beer festival.

### August 4th 2012

There's always been a part of me that wishes I'd owned a horse. Instead of diminishing with time the feeling seems to grow and I find myself looking at horses more and more, with the knowledge that, somehow, I've missed out on something. And I probably have. There's a sort of horsey social life that I find difficult to define. I've heard conversations like, 'There's new people moved into that little farm down the lane.' 'Do they kedep horses?' And if they do, they seem to leapfrog into a social circle that seems closed to people like me. People at the top of the local social scale all have horses, they are all friendly with me, but only up to a point – I don't have the social icing on the cake that a horse would bring.

Those people who have moved into that little farm down the lane do keep horses, fairly nondescript horses. They travel about to hunt meets and farm rides in a trailer and vehicle that together are worth about £500, they look to me as if they are on their financial arse, but they are greeted effusively by all and sundry, high and low, and it's all kissy kissy and I've never done better than a cool handshake.

The women that are involved with horses are either very beautiful or look like their horses, but stories filter back of escapades at hunt balls that only add to the feeling that you are missing out on something. To say I've never owned a horse is not strictly true. I've owned three but they weren't really for me so they don't count. An old man who used to work here had worked with horses all his life. 'You need to get those children some ponies,' he used to say.

I didn't need much encouragement and we went off to a horse sale one Saturday and bought two Section A foals that had come off the mountain that morning. Wild! We used to have a walled kitchen garden (someone was here the other day and remarked 'this must have been a gentleman's residence at some time.' 'Still is,' I said), so we put them in there and I'm sure they could climb the walls several feet off the ground. But Bill did break them, sort of, and I used to ride them, sort of, but it was more a triumph of my weight and strength over little ponies.

The trouble with having ponies was that someone's children always wanted to ride them and one day a friend's child who said he was a good rider got dragged and skinned and I decided that that was enough and if we didn't have ponies, no one would want to ride them. So I swapped them for two donkeys. Donkeys don't do anything to get you up the social scale and certainly don't get you kissy kissy with the toffs. But they were safe.

One thing leads to another and within a few years we had five donkeys, which is close to being overrun with donkeys, so they also had to go, but it wasn't that easy to move them on and I had to take a Shetland pony colt in part exchange. He was a nasty little sod and spent his time eating or chasing sheep. He would get a ewe down by chasing it until it dropped exhausted and then he would kick and bite it until he got bored. Can't remember what happened to him and I don't much care.

It is still in my mind that, should I lose my driving licence,

I would buy a Shire horse and cart to go to the pub. I look at horsey adverts and note the adverts for carts and carriages. There's an eccentricity to it that appeals to me. The horsey people I know tell me the horse trade is dire, anything any good is dear enough, but anything plain and ordinary has little value. This puts an ordinary horse and horse box well within my reach. I already have the very ordinary truck to pull it all. It's a tempting scenario, I could easily make that great social leap, but would I want all that horsey hassle? While I still have my driving licence, I don't. A friend, (a horsey friend) was telling me she went to a horse sale recently. There were three lots of mare and foal. At the end of the day they found that the mares had all gone with their new owners, but the foals had all been left behind. Just how sad is that? When animals lose their value, terrible things happen to them.

## AUGUST 11TH 2012

'Is that your field up there?' It's a walker and by his tone I'm on full alert. The tone tells me quite clearly that he isn't about to tell me that it's the best crop of winter wheat he's ever seen. My mind tells me that he's found something wrong. Time for quick thought and before he can go on, I've guessed the problem. It's a big field and there's a public footpath through it. I'm supposed to spray out a metre wide strip through the wheat for walkers but I haven't. I don't actually like spraying out a barren strip and usually take the lawn mower up there and cut the path out. But it's been a wet summer and by the time it was dry enough to cut the path out, the wheat was too long. 'You're supposed to spray a path out for walkers. Look at me, my legs are soaked through.'

I've had this sort of conversation before (as you can imagine), I got myself off the hook on a previous occasion by saying there were curlews nesting there and I didn't want to disturb them. I could try that again but I need to challenge myself to come up with something new. 'I'm very sorry you got wet but I've decided I'm not going to clear any more footpaths until I get a price for my milk that exceeds the cost of production.' 'Don't blame you,' he says, and he gives himself the sort of shake that a Labrador would give itself to get water out of his coat, bids me a friendly farewell and sets cheerfully off through a field of kale that doesn't have a footpath cleared either. Seems I'm turning into a bit of a chancer.

***

Life can be a roller coaster for all of us. If you are a turkey, it's no different. If you are a turkey you can have your highs and your lows, your good days and your bad days and in the background to your life there is always this thing that the humans call Christmas. Since the demise of Neville our cockerel, Eric the turkey has blossomed into a great character. He originally had three wives. Two of them became ready meals for foxes and his remaining wife is busy sitting a clutch of eggs (5 days to go) and so Eric has sought company with humans and any reflections he can find of himself in clean cars (not ours) and any glass that is of his height. We were mending a feed barrier the other day and Eric was in the thick of it, back and fore fetching timber, nails and tools. He's discovered his reflection in the windows of what we call the dairy, an old larder next to our kitchen, so he spends a lot of time around our kitchen door, which is where we find him today.

It's 9am on a Sunday morning and two ladies are preparing breakfasts for bed and breakfast guests. As much as possible is prepared half an hour in advance and so when the guests appear it all gets a bit frenetic as kettles are boiled, toasters crammed and huge 'full English's' are assembled on plates. How they eat all this food is beyond me, given the state of some of them when they came in the pub last night. 'Fried bread?' 'Yes please.' 'Mushrooms?' 'Yes please.' 'Sausages?' 'Yes please.' The list is endless and it's not a list that is ever offered to me.

The two ladies concerned get a bit edgy as they pile food on plates. Fried bread, I notice, can have a mind of its own and will slide off a plate unless it has an egg or two on it to hold it down. The edginess is not helped by the fact that they have to keep the kitchen door open. If the door is closed the toaster sets off the fire alarms so we have to keep it open. In the doorway stands Eric. He is also waiting for some breakfast – he has developed a liking for toast as well. Annoyed glances go from the women towards the turkey and unladylike mutterings are just discernible.

Because Eric is not just standing there, he is on full display. His feathers are all fluffed out and he makes little strutting movements. He wattles are bright crimson (I think they call them wattles), they are so bright red you would think a heart attack was imminent. But that's not the half of it. Turkeys say 'gobble gobble' but it doesn't do the noise Eric is making any justice. The noise he is directing through the kitchen door is loud and endless, it is making the guests in the next room laugh and the ladies getting their breakfast really twitchy. I'm watching this scene unfold and the annoyance is starting to be directed towards me.

Time to go now, but I'm not quite finished. I squeeze out of the door past Eric, who reacts with even more noise and it's a simple matter to nudge him through the door with my foot and to shut the door behind him. The noise of the women escalates to match that of the turkey. Me, I'm off up the yard to safety. That was Sunday. There's no sign of Eric on Wednesday although we look everywhere for him. We find him on Thursday drowned in a water tank where he would sometimes look at his reflection in the water. Turkeys can have good days and bad days and for Eric this was a very bad day. That's Neville and Eric I've lost this year. It's not easy being a dairy farmer.

## August 25th 2012

I've got this 15-year-old grandson and I'm teaching him to drive a tractor. He reminds me of my father in some of his ways – he thought he knew everything as well. We are carting bales of silage and he is driving and I'm sitting on the little folding seat that they fit on modern tractors. He's struggling to find the right balance between clutch and accelerator every time we start off. We all did, I suppose, and to be fair he's got 150 horsepower under the bonnet and a heavy load behind. Every time we move on we either have a dramatic stall or a dramatic jerk, because he has too many revs on. I impress on him the need to do the start-off smoothly but he's not that concerned. 'We're moving aren't we?' It's OK for him, I've already cut my head twice on the roof of the cab as he's clattered me against it. I try to explain to him how different it was for me when I first drove a tractor in a harvest field because there were always men working on the load stacking bales or sheaves and if you started off with the sort of jerk he was achieving, they would be thrown to the ground.

He's rolling his eyes at me when I'm telling him this so I tell him that if you throw someone off a load, or even come close to it, you were likely to get a sharp smack around the ear. Just to prove the point I give him a smart smack around the ear and he stops rolling his eyes at me.

When I was a boy we lads were always hanging around farms in the village at harvest time. We helped as best we could, we didn't get paid, but if we were any good, we were fed. We all enjoyed it and wouldn't have been anywhere else. The first tractor I drove in a harvest field was a Standard Fordson; you see these at shows still. It had a clutch and brake combined, you pressed the clutch down, and then it activated the brake.

I was chosen out of all the other boys because I was the only one heavy enough to press the clutch down and I could only do this if I stood on it with both feet. Without being too technical, you can just imagine how tricky it was to start off smoothly when it took your whole weight to push the clutch down and you had to start removing some of that weight a bit at a time. That was why it was just as important to automatically shout out 'Hold tight!' every time you prepared to move on. It gave the men on the load a warning to steady themselves as you prepared to move.

The last time I shouted out 'Hold tight!' to anyone on a load was the year my son came home to work after college. We were feeding cows stock feed potatoes while they were out at grass. I was driving the tractor and he was shovelling the potatoes off the back, which was how it should be. Every time I moved on I shouted out 'Hold tight!' and he looked at me as if I were mad. After a while I stopped doing it because I could see he didn't like it. I've never done it since, never needed to in fact, so 'hold tight' finished that day in our family in that field, feeding those cows potatoes. As we went on I wondered at the difference between my son on the load and those men working on my first load. The men were wearing shirts with no collars, corduroy trousers and hob nailed boots. My son was wearing trainers, jeans with holes at the knees and a white T-shirt. The T-shirt had a cartoon on it showing a young man and a young woman in a very popular horizontal position and under it the words 'The Welsh Agricultural College idea of a short term ley.'

Huge changes have taken place within the timescale I have alluded to, but the basics are still the same. There's still a need for young boys to get a tractor into motion smoothly.

### September 1st 2012

I lived my early years in a town and my Dad had an allotment. Some allotment holders were members of a pig club. My Dad was in the pig club. I can't remember how many pigs they reared but I do remember being in awe of these pigs as a small child. I was scared of the noise they made at feeding time and the big pigs due for slaughter were simply monsters.

If you were in the pig club there was a duty rota for the members that involved a week's cleaning out and feeding and the feeding involved cooking pig swill. The pig swill had to be collected and there were 'pig bins' at intervals all around the immediate area. We are talking about immediate post WWII here, where everything was still scarce and waste of any sort was not to be contemplated. It just shows how perceptions change: it was considered acceptable at the time to have these pig bins scattered about the neighbourhood, for people to put their scraps in and boy did they stink in the summer.

They were all collected up every weekend and an empty

one put in their place. It all had to be cooked in a boiler (before it was tipped into a big bath and was fed to the pigs) and some meal mixed with it. The meal was so precious it was measured out as if it were gold dust. One of the buzz-words in the farming industry at the moment is 'sustainability': using resources wisely, not relying on resources that could become depleted or wasting resources to move them about. Just imagine how many pigs you could rear if you built a piggery at the back of a supermarket and fed them the waste that surely comes out there? Just look at all the food that goes back to the kitchen on plates in restaurants and pubs!

As a child I can remember an alcove in my bedroom with a side of bacon and a ham hanging suspended on a broomstick that was lodged on the picture rail. I didn't like them being there but you didn't dare complain. One night the broomstick broke and they came crashing down. I thought the end of the world had come. The bacon and the ham were the rewards for taking the trouble to collect, cook and feed pig swill not to mention a share of the pig muck that went back onto the allotment to grow the vegetables to go with them. That was sustainability in a real form.

So what triggered thoughts of pig swill? Well I've been thinking about pigs for a long time. I like having a pig. You can talk to a pig. At the moment I don't have anywhere to keep a pig but I could fix that. One evening I went to the pub and they were short-handed and had about 80 in for a meal, so I ended up washing up. There was so much food wasted on plates that I thought, 'I could keep a sow on all this.' I'm fairly sure it's illegal to feed pig swill now but I wasn't going to tell anyone. When I eventually went for a drink I mentioned food waste and that started a conversation off that had a sequence to it so that in the end someone tells a story about pig swill.

Many years ago, this man and his brother had gone to plough and sow a field for an old lady that lived alone on the next farm. She kept pigs and fed them on swill. The pigs were the great delight in her life and she had two rows of pre-cast concrete pig sties. These sties had had their day and one or two of the roofs had fallen in, but at an angle against the wall, so you could still keep a couple of pigs in there. All morning they watched her go up and down, cleaning out or carrying buckets of steaming swill to the noisy pigs. There was pig muck and pig swill everywhere, all over her, and she wore a chin-to-floor apron made out of a sack that she had fairly obviously been wearing for some time. At midday she stops work and tells the brothers to come in for some dinner. They are not sure about his, but they don't like to say no, there's something of the witch about her.

So they go into the kitchen with her and there's a stew bubbling on the stove and it smells fine. There are some used dinner plates on the table that she gets off and washes in the sink, all the while chatting away contentedly. She puts the dishes she washes on the draining board and when she's finished she dries them on the sack apron she's been wearing all morning attending to the pigs.

The two brothers haven't missed any of this and are getting really worried now, their imaginations working overtime but too scared to say anything. She puts two of the plates on the table and ladles some stew on. Then she gets a loaf out of the pantry and cuts some thick slices off it. But only after she's given the carving knife a wipe with the same apron.

So the two brothers are sitting there staring at these plate-fuls of stew and wondering what to do next but definitely too scared to make any complaint. 'Get that in you, it's a cold day, there's plenty more, I'll make you a cup of tea to go with it.' We can only begin to imagine the thoughts racing through their minds as they connect the thoughts of pigs, pig muck, pig swill and apron. One brother decides that there is only one way out of this dilemma. He closes his mind to these thoughts and wolfs down the food as quickly as he can. The other brother is made of less sterner stuff. He waits until the old lady's back is turned and very deftly swaps his full plate for the empty one of his brother. She turns back to the table, 'Come on, eat up, your brother has finished his.'

## SEPTEMBER 11TH 2012

We've just done our third cut silage. The Keeper always wants to know when we are cutting the grass. The mower leaves more than just grass behind it. There's all sorts of things get left behind the mower: dead mice, rats, hopefully a mole or two, probably the odd young rabbit, probably a few hen pheasants too slow to get out of the way. In the daytime this food source brings out the birds of prey and the scavengers but at night it brings out the foxes.

The Keeper has his new pheasant poults out in his pens now and if there are foxes about, he wants to know about it. So he goes out late at night with his lamps and his rifle. I ask him next day how many foxes he has seen and it's only one and that won't trouble him again. I also ask how many hares he saw. 'There were 17 on your field on the bank. They were all chasing each other about and we sat and watched them for ten minutes, it was a wonderful sight. The field above was also full of hares, we gave up trying to count them.' This news just delights me, but it's a bit scary as well. Contractors and walkers see them, and tell others, and that is a bit of a worry.

\*\*\*

Last week I was asked to take a neighbour's daughter to church to her wedding and I decide to make a good job of it, so I go down to my Eastern European friends at the car wash, and order a full valet. They think this is good. They have the price options on a big board, car, van, 4x4. When they open the door I can see they wish they had an option that included the interiors of farmers' cars – they don't like the look of it at all. They speak good English, well I do anyway, and we eventually arrive at, 'Boss, he wedding.' And I can see it motivates them to give the car their best shot.

When they finished it looks like new and they stand back to admire their handiwork. 'Very fine car.' So I end up explaining how I bought it and how much it cost and they try to buy it off me. The full valet cost £60 and I can see them counting the money in the mirror, they are obviously disappointed that there isn't a tip, well it isn't my daughter that's getting married. I drive carefully home and manage to get past my neighbour's before his cows cross the road. Next day there are torrential downpours. Someone calls to say that the road at the bottom of our lane is flooded with muddy water that has followed the wheel marks off a wheat field where they are carting bales. So I have to have a look and decide that if I drove through at walking pace it wouldn't be too bad but if I met someone coming the other way at speed, the car would be covered with muddy water, never mind the white ribbons. Stephen who works here is very thoughtful and he's worked it out as well. So at the due time I drive carefully through the flood and up to the top of the hill where the wedding takes place. Stephen has bought drums of water and sponges and we wash the car off and then fix the ribbons. All this is done in another torrential downpour and it's a very wet chauffeur that eventually opens the Jag doors for his first load, the bride's mother and the two bridesmaids.

\*\*\*

I've got this hen turkey with two little poults living within the safety of an old stock trailer. The trailer is parked up by the calf pens. I said to Stephen last week, 'I'll let that turkey out now to run loose.' 'I wouldn't if I were you, there's a cat up there in the bales with two kittens, she'll kill those two chicks in no time.' This isn't what I wanted for my turkeys. I didn't want them confined. I wanted a nice little flock of free-range turkeys that came to the kitchen door for toast and roamed at will in safety.

All my turkeys are now confined. The big ones to protect them from the fox and now the very little ones to protect them from the cats. Yesterday I was working up by the calves cleaning the pen out. There was no sign of the cat. I thought, I'll let her out while I'm up here and can keep an eye on them. I let the tailboard down and the mother turkey and one poult were very soon out and delighted to be so, picking at grass and scratching about in the straw. The little baby turkey had only been out about 30 seconds when from under a pile of pallets come two wild kittens. They are very small themselves but they are down on their bellies stalking the turkey in the classic stalking mode that you see with wild cats in Africa. I soon had the turkeys back in the trailer and the ramp safely shut. That's no good. Have to leave them another couple of weeks.

## SEPTEMBER 15TH 2012

We've got the pond, it's a feature of the field in front of our house, and I'm quite content with it as it is. My wife is not content with it (no surprises there then). She's always wanted swans on it. Twice in the past I have sourced a pair of swans from animal rescue centres. If you can contrive the arrival of the swans with her birthday you are talking multiple bonus points. But swan rescue is all about rehabilitation and after a few weeks or a few months, depending on why they were rescued in the first place, the swans will fly away, somewhere else. From the point of view of the rescue centre this is very good. It is a success story, it's what they are all about. For my wife, with her swanless pond, this is very bad.

Naturally this is my fault. But I never give up on anything and for this year's birthday I secured for her, not one swan, not a pair, but a whole family. Mum, dad and six cygnets! You may think this is a bit over the top but it's what was available. Of course, when the cygnets fledge they'll fly away or be driven away by their parents and I doubt if the parents will stay anyway but, like life itself, you just go with the flow and enjoy the good bits while you can. The only dilemma I can see on my horizon is whether to have swan or turkey for Christmas dinner.

\*\*\*

I've never known how old my dog Mert really is but there's more grey around his muzzle than there used to be. He's still a willing worker but you have to call him, whereas two years ago he was always with you wherever you were. The biggest change came in his behaviour after he had an operation of a very private nature, which stopped the arrival of endless litters of corgi cross sheepdog pups. After the operation he became less active, and inevitably put weight on. This concerns me because quite simply I couldn't manage without him. His presence is essential when I'm feeding or working with cattle. I was cutting thistles one afternoon and when I'd finished, the heifers in the field gathered about me at the gate. Mert wasn't with me because there's no room for him on the tractor I was using and there was no way I could get the tractor onto the road without the heifers following me out of the field. If Mert had been there it wouldn't have been a problem, but as it was, it was 20 minutes before the cattle got bored and wandered off. I've even started thinking about looking for a replacement but as anyone who has ever had a working dog knows, it isn't that easy because if you have an exceptional dog, it might be some time before you get another. It seems logical, in the short term, to get Mert fit again.

So when I go around the cattle every morning, I get him out of the truck when I'm on the very top ground and make him run back down to the road, which is about a mile. He copes with this quite well and as he keeps alongside me he looks up and his expression says that he thinks that I should run down from the top on alternate days but until he learns to drive, this is how it's going to be.

Next I turn my attention to his coat. It's become very thick and matted. Over the years I've had bearded collies that have needed shearing right off every year but I've never had a border collie clipped. His thick coat is obviously a burden to him so I book him in to be trimmed. This turns out to be a three hour adventure. Mert fights the process every step of the way. After five minutes we decide he will need to be muzzled. He hates being clipped, but not as much as he hates being bathed. After the shampoo he smells like a teenage boy that has been given too much aftershave for Christmas but it's after the shampoo that removes years of dirt and dust, that the clippers are able to make real inroads into his coat. I'm very impressed by the expertise of the whole process and am amazed at how much coat is actually removed. I ask the lady if she will cut my hair in future and she says she's willing. I ask her how much it will cost and she says it will be £28 the same as the dog. I tell her that I only pay £4.95 at the hairdressers I go to. 'Yes,' she says, 'but does she do your anal glands?' Think I'll leave arrangements as they are. Mert is transformed, it takes years off his demeanour, he's more active, busier about the yard and looking for work. Even if he does smell like a tart.

### SEPTEMBER 22ND 2012

This man comes up to me at the Dairy Event and says he once bought a pup off me. I can't remember the man or the pup so we piece the story together. I have to write this next bit quietly because Mert is just outside the kitchen door. He knows I'm writing about him because he gives his tail a 'flip'. I once had a bearded collie bitch who was the best working dog I ever had. Part of a successful business is to have a succession policy so I bred a litter of pups from her so I could keep a bitch pup to carry on the line. It was from this litter that the man's dog came. The bitch I kept developed a tumour at two years old and that was the end of that line for me. But the man at the show tells me his dog is the best dog he's ever had and that he had taken him to breed a litter with someone else's bearded bitch so he could continue the line. He had a bitch out of that litter that is now about three and she is a remarkable worker as well. This is of great interest to me and I arrange that should his bitch have a litter of pups, he will save one for me. I'm quite chuffed to think I might get back to that breeding when I thought it was lost to me. He calls his bitch Poppy. That was what we called our bearded collie bitch! What a coincidence! It's a nice story with hopefully a happy ending.

\*\*\*

The sheep farmers are always on about the numbers of lambs they get. We've had five sets of twin calves over the last month. This isn't necessarily the nature's bounty it might appear. If the calves are both sexes, the heifers will not breed and we call them freemartins. So you end up with two small calves that aren't much use to you. In three cases the calves had got tangled up together and the birth had not proceeded: we had to sort the calves out in the womb and in each case the calves were born dead and the cows had had a difficult time. A nice healthy single calf, born naturally, is to be preferred.

***

So it's a beautiful sunny Monday morning. In the background, the beautiful rolling hills, a feature of where we live, stretch into the distance. In the foreground is the field in front of our house, nicely treed (if there is such a word) and our pond is the centrepiece of the scene. And there on the pond like stately galleons are our newly acquired swan family. A friend collects me to go to a meeting, and as we drive across the front of the house, he stops, he notes the scene, as I have just described it to you. He particularly notes the swans and how beautiful they look on the pond and I tell him how I acquired them for my wife for her birthday.

Now it's a beautiful Monday evening. In the distance, the sun is dropping down towards the distant hills. In the foreground is our pond. It's swanless, as is the field. If you want to see our swans you have to lift your eyes back up in to middle distance. There, about three fields away, not our fields, is a larger pond and I can see the swans there. When I get my binoculars out of the house I can count them and they are all there. It's a bit of a mystery because the cygnets can't fly, but the story unfolds. The swans got through the hedge, walked 400 yards down the road, crossed the three fields and found a new home. From a swan's point of view this is a good story, back to the wild. They probably don't know yet but over the next wood are two lakes, each nearly a mile long, and that's where they will probably end up. From my wife's point of view, her birthday present lasted three days. Naturally, this is my fault.

**SEPTEMBER 29TH 2012**

We leave 20 acres of stubble every year for the birds to over-winter on. It's a part of the environmental scheme that the landlord runs. But just because you don't plant a crop, doesn't mean that you don't get one. I have to leave this field for about 12 months or until there is another stubble somewhere else to replace it. So in the spring it grows a crop of weeds. Seeds can lie dormant in the soil for years and years and so up they pop and flourish. It's mostly weeds, grasses and thistles and by mid-summer it's a farmer's nightmare. The thistles go to seed, and with a fair wind, populate the parish with thistle seed for future generations.

This year I was allowed to spray the field before the thistles went to seed. 'They' that design and administer these schemes take years to react to the negatives that they create! So 12 months after I did this in another field, I am trying to chop up this mass of dead vegetation. There are no birds here because I suspect it's difficult to take off and land, another bit 'they' hadn't thought through, but it's good cover for a fox or a hare. It's not long before I see the first hare, but she doesn't leg it to the next field, she hangs about. This tells me there's a leveret here. And there is, and I soon see it. It's as tiny as a baby kitten and I see it several times as I progress across the field with my work and mother hare is never far away. I never do this work in one day, I make a point of doing it over two days, which gives the wildlife a chance to move on somewhere else. There's no sign of mother or leveret next day so that's good. And I wonder what chance this tiny leveret has with so many predators about and with winter fast approaching.

**OCTOBER 6TH 2012**

I've always been an advocate of having a succession policy. People come and people go and there is an inevitability about that, which demands that you are able to move on seamlessly from one to the next. Thus it is also with animals so it is a source of much regret that I took my eye off the ball with succession. While I was able to take so much pleasure from the notoriety of Neville our cockerel,

I hadn't considered what would happen when the fox took him for his dinner. So we have to go back to the start and find out where we went wrong. We turn baby day-old chicks into pullets that go on to populate free-range laying units. Sometimes, by some fluke of nature, we get two or three cockerels amongst nearly 40,000 pullets. That is where Neville came from – he wasn't wanted with the pullets so we simply let him out into the yard and in a short time he became the character that some loved and some feared. The next crop of pullets were destined to go on to produce eggs that would be used in the production of vaccines and, for reasons that I don't understand, these eggs have to be fertile.

So there were 4,000 cockerels amongst them. Surely we would find a new Neville amongst them? So the pullets and the cockerels move on, but the cockerels have flourished, there are too many cockerels, and we are left with 100! This is a succession policy that is out of control. What do you do with 100 cockerels, especially if they are lightweight birds, a strain bred to lay eggs and not produce table birds?

Well, first of all you look for a new Neville and as you are determined not to be caught out again, you let eight would-be Nevilles out on the yard. And you try to sell the rest, a few at a time. But it's not easy and you need to sell them more quickly. So your son puts 30 into poultry crates and takes them to a live poultry auction 30 miles away. There are £ signs in his eyes as he sets off with this bounty. Dick Whittington didn't set off for London in better spirits.

Four hours later he returns with an empty fuel tank and £9. I've told this story in the pub. '£9, that's a good price, I didn't think you'd get as much as that for them.' And I had to explain to them, as I have to you, that it was in fact £9 for the entire 30. Strangely this price has stimulated some interest and we've sold small numbers locally at 30p each, embracing at the same time the modern day marketing strategy of 'buy one get one free.'

So we're down to about 25 now, plus the eight would-be Nevilles. None of these have attacked anyone yet, which is disappointing. We were carting straw bales the other day, I wasn't driving the machine that was unloading the bales so I sat down on a cattle trough that was nearby. Slowly, one by one, the eight free-range cockerels came across and sat in a row next to me. A friendly gesture? No, they were quite sinister, even threatening. I was very pleased.

\*\*\*

There's a natural course to events in people's lives. A natural sequence. Take my eldest grandson. He has his first hangover. A couple of weeks later he's sick all night and most of the next day and vows he'll never drink alcohol again. Then he's seventeen and gets his driving licence, closely followed by his first car. (A Corsa, Corsas are cool). Then he passes his driving test and then gets his first proper girlfriend. We see her about quite a lot at first then two weeks go by and I've not seen her. So I ask him if he's still courting. 'What's that mean?'

## OCTOBER 13TH 2012

Today I'm going to tell you about sheep. Not just any old sheep: Welsh Mountain sheep. I don't keep sheep any more but love and admire them from a distance and Welsh ewes are my favourite. They are hardy (they have to be), good milkers (feed their lambs well) and are really excellent mothers. They are a wonderful ewe to cross with a downland ram to produce cross-bred ewes that still retain the milking and mothering ability of their mothers but with better carcases than pure-bred Welsh and with better proficiency (more lambs). Welsh ewes vary within the breed, smaller ewes in the North down into Mid-Wales and a different type of ewe on the Brecon Beacons which has a lot of Cheviot blood in them from Scotland.

And then we have the South Wales Welsh ewe and it is these we concentrate on today. They are a bigger ewe than those that live to the North of them. Some of them will have tan, sandy faces and necks and are called Nelson ewes. Nelson is a village

in Glamorgan, just south of Merthyr. A friend of mine says 'I'm
going to buy 100 Nelson ewes this year, you fancy coming for a
ride?' You bet I do. I wonder to myself if he knows what he's doing
because some of these ewes manifest their mothering and milking
abilities in a search for food that will take them over fences,
under fences, over gates, over walls – still, that's his problem. In
describing them I find myself comparing them to big cats. They
prowl about looking for opportunities for escape like caged tigers.
There's something of the leopard in them. I've never seen one
climb a tree but they can climb most other things and I'm sure they
could and would climb trees if trees grew grass instead of leaves.
And there's something of the lion in them because lots of them
have the tan sandy face extending down their necks, like a mane.

When the auctioneer starts off, he tells purchasers to
follow their sheep back to their pens immediately and to put a
mark on them. My friend is surprised at this, he tells me that
people gathered there look decent tidy people, 'They don't look
the sort of people to steal each other's sheep'. I tell him that I don't
think it's stealing they are worried about, it's the sheep jumping
from pen to pen and getting mixed up.

I just love it all, a sheep sale, and have to keep my hands
firmly in my pockets. My friend only buys a few because he can't
locate a lorry with room for more. I get into sheepy conversations
with sheepy people: there are ewes there from the Rhondda valley
and the valleys of Monmouthshire. One disappointed vendor has
failed to sell his sheep. They were only offered £100 apiece and he
wanted a lot more. 'I could have had £120 for these in Builth last
week.' I've come across this phenomenon in sheep people before. I
hear them in the pub. 'I sold my lambs for £80 today, it wasn't very
good. I could have got £83 for them last week in Welshpool.' They
genuinely feel aggrieved, cheated almost. I find it very strange. I
don't know about you, but in my experience, last week is gone,
and as far as I can make out, gone forever.

So we drive home from the sheep sale on a beautiful sunny
afternoon and my eyes are on the dramatic scenery of mid-Wales

and the beauty of it all. But my mind is on sheep. If I had another 100 acres of land and it was well fenced I would have bought five or six hundred of these Nelson ewes and because I would be taking them onto lower land away from the harsh environment of the mountains they would have 150% lambs, that would be 800 lambs at £80, that's £72,000. It looks really good. Mind, I would need a couple of good boys to catch them if they needed catching. Perhaps my time for 5-600 Nelson ewes has long gone and I can but dream. My friend phones this afternoon and says he's going to Nelson again on Friday and will I go with him? I don't hesitate. I visualise the sale field again as we left, all the hustle and bustle as sheep are counted and loaded, the shouts of the men and the bleating of the sheep. And in the distance are three ewes that have managed to escape from the pens, nibbling away at the grass as they make their leisurely way back up the mountain.

*** 

We have what we laughingly call a committee meeting on Sunday nights in the pub, the Sunday evening following our midweek diners' club. It's when we decide what we will have to eat next time. I am sure that those not present think it's a formal occasion but what really happens is that there is a lot of reminiscing about what we used to eat years ago until someone happens to mention something we all fancy and that's what we choose. It's also an occasion to wind up the landlady by suggesting something like road kill or squirrel or badger (road-kill badger). This usually brings out the riposte: 'If you think I'm cooking that you can think again.'

Someone also suggests 'beastings pudding'. People used to eat this years ago. It is made from the fore milk or colostrum that a newly-calved cow gives, which is thick and yellow. As far as I remember you just heated it up and the result was something that looked like egg custard. Beastings pudding also comes into the landlady's 'if you think I'm cooking that' category, but that doesn't necessarily stop the discussion. We are advised that there is nothing like 'a good gutsful of beastings pudding' followed by a pint of cider. We are reliably informed that this causes the pudding to curdle and you can feel the lumps bumping about in your guts. There were two newcomers to the committee last time. I'm supposed to be chairman. I let them choose what they like, which is how good chairmen operate, though we've never had anything I don't like which is also how good chairmen operate. The two newcomers ended up choosing the next menu, sausage and mash and onion gravy followed by summer pudding with tinned evaporated milk. They influenced the chairman by blowing kisses and fluttering eyelashes. And that was just the husband, I'm not telling you what the wife did!

## OCTOBER 20TH 2012

I'm getting twitchy about the weather. It's October and we have 20 acres of spring wheat to combine and about 50 acres of grass to mow. And it's making me twitchy. We did an hour at the wheat on Sunday (it was 25 acres then). It was about 19% moisture on a dull day, which means that if we could have a couple of sunny days, preferably consecutively, we could get it at around 16% moisture which is more manageable. My top man is away on holiday next week so on a wet day we abandoned wheaty thoughts and turned our attention to getting next year's crop in.

A day later we return to the present wheat crop. It's the field that we will leave in stubble for 12 months for the birds. I know that in 12 months' time it will be the jungle of weeds and grass that I moan about periodically but in the meantime we have to cut it fairly high to avoid getting any weed seeds and grass into the harvested wheat and spoiling the sample. This is leaving a nice carpet of weed grasses that have seeded.

This is my 1,000 feet field and it's desperately hard up here in the winter. There are a good number of skylarks that live here and goodness knows how they survive the winter. The stubble and grasses I am leaving will provide food and shelter for them as it will also for the hares that live up here. I know I will be moaning when I have to do something with this field next year, but it all seems worthwhile if it is helping the wildlife through the winter. Anyway I'm quite good at moaning! It's what we farmers usually do. If there's a shortage of milk by Christmas you'll know we don't always moan for no reason.

***

I read lots of adverts. I sometimes wonder if it's a compulsion. I like reading mostly farming adverts and occasionally car adverts. When I become Prime Minister one of my first pieces of legislation will be to ban people placing adverts that don't also carry a price. It will put an end to all that conjecture, 'I wonder how much they want for that?' The weekly local and farming papers carry rich pickings for me and it was in our weekly local paper that I was reading notice of a farm sale. When someone retires from farming or ceases farming, they often have what is called a dispersal sale. They usually sell everything they have and that usually includes what can be best described as a lot of old tat. I've been to sales and I can't believe what people will put out to sell. Even more unbelievable is the fact that there are people about that will buy it. I sometimes wonder if it's sold because it costs money to take it all to a tip.

Anyway I'm reading this list of goods and chattels and livestock and machinery at a dispersal sale to be held next week at a farm an hour away, which I have no intention of attending. It's mostly sheep and sheep stuff but I read on. Then I come to an item that I read twice. 'Black and white Welsh Collie bitch (9 years).' I can't believe it. The adverts tells you quite a lot more than that. With ewes and lambs there are around 800 sheep at the sale. You can't manage all those sheep without a good dog. How do I know it's a good dog? Because it's nine years old. They wouldn't have kept it nine years if it wasn't a good dog.

So we have this nine-year-old bitch that has served this farm faithfully for nine years, rain and shine, frost and snow. And what do they do with it when they decide to retire? They say thank you by putting it up for sale with their pots and pans, bits and pieces and their tat. They don't know who will buy it, what sort of home it will have, and presumably don't care. Just how callous is that? I bet the farmer has bought himself a nice cosy little house in the village. I bet the bitch thought she had done enough to have a bit of that as well. There's a bit of me thinks I should go and buy her, or at least have a look at who is bidding on her just to see if I like the look of them. So I'm a bit of a softy. I don't care. Last night in the pub one of the farmers says 'Did you see the advert for that farm sale next week?' 'Why?' 'They're only selling a nine-year-old bitch, miserable sods!'

\*\*\*

Sheepy footnote: I recently told you about my visit to a sheep sale in South Wales and in particular I told you about the South Wales mountain ewes found in that area, known as Nelson ewes. A part of that story was the reputation these ewes have for escaping from just almost wherever you might put them. On Sunday I turned the TV to *Cefn Glwad*, a programme of the Welsh language channel S4C. Not Welsh speaking me, but I think it means supporting the land or countryside and it's a programme I like very much because it mostly visits farms. If you like seeing other people's farms it is excellent and you can enjoy it because there are English subtitles available (it's 134). On Sunday they were visiting a farm in the Rhymney Valley. I used to work on a farm that had the Rhymney river running through it. 50 years ago the river ran black with coal dust. There was no life in the river and no life on its banks. These days you can find trout and salmon in it. I recognised the lady who ran the farm immediately as one of the vendors I had seen at the Nelson sheep sale. The conversation between her and the programme host, Dai Jones, turned to Nelson ewes and their nomadic tendencies. Dai Jones says 'They reckon that if you have one in the freezer, you don't want to keep the lid open too long.'

## OCTOBER 27TH 2012

I've just stopped and examined my life. I wonder if I'm slipping away into some sort of revisited childhood. My first job, as I go out after breakfast, is to feed my turkeys. They live close to the house in what used to be a walled kitchen garden. There's eight adults and two babies and I've just noted I've started talking to them. Turkeys like a chat. They gather round you making sort of chirruping sounds, but while you are speaking to them, they all fall silent. Then when you stop they all do their gobble gobble bit, that's until you speak to them again when they listen to you again. So I find myself saying 'Good morning' and 'Is that enough wheat or did you want some more?' You can have a good conversation with turkeys.

Then I go to feed George. George is a homing pigeon who's lost his way. He lost his way about a month ago now and I feed him twice a day. I can't always see him when I go to where I feed him, but he's always there in seconds. If it's a wet day he's on the roof trusses of the shed we laughingly call a workshop. Stephen who works with us is not best pleased because what comes out of pigeons' back ends lands on his spanners. Then yesterday another homing pigeon turned up. It's a lot bigger than George, which makes me wonder if George isn't a George after all. Two pigeons, apparently, can make twice the mess on spanners as one.

Next stop, the kittens. We've got an implement parked next to our bays of straw. There's four kittens been brought up there and they are quite big now. We rarely see these kittens until they are very mobile by which time they are unapproachable balls of spitting, biting aggression. When I sneak up on them they disappear either under the machine where they were born or, more likely, into the gaps in the bales, where they are safe and warm. I put them some calf milk powder down every day, you can't feed them in a more conventional way, I just put a cupful of powder down in a dry place and next day its gone.

Next, the cockerels: there's 20 left in the trailer and eight roaming the yard training to be Nevilles. Some cockerels have been taken lately by gamekeepers to put with their pheasants. The theory is that the pheasant will follow the cockerels about all day and not stray, while the cockerels for their part don't go far from their food source, so you can see how it works. Yesterday we were spreading manure. I was driving the loader tractor and Stephen was the spreader. While he was spreading a load, I switched off my engine and took a look about me. There's a little wooded valley below me. The Keeper has three different pheasant pens down there. A cockerel crows loudly amongst the trees, two others answer at various distances and it's a strange sound in that environment. Most nights in the pub people ask how George is or what the turkeys have got to say. Reckon they are as daft as me. It's that sale tomorrow where they sell that nine-year-old sheep dog I told you about. Everyone reckons I should go and buy it.

***

What a difference a couple of days make. I write these notes a bit at a time as my life unfolds or as I lurch from crisis to crisis, depending on your viewpoint. So two days after I write 'no combining this week then', I have a lunchtime snooze and wake up and it's been a very wet night but now the sun is shining and there's a gale blowing and I look out of the window and, I wonder. I wonder what it's like up on the top in this weather? So I phone my nephew, who does our combining, to ask what he thinks. He's keen to try it. Partly because he wants to get it done so he can put his combine away and partly because his wife has just told him to clean windows.

    Within an hour we are away, and it's OK, in fact it is quite good for mid-October. We keep on until nearly nine o'clock at night to get it finished and it's such a relief to get that wheat in. I was talking to my consultant just a day or so ago and he's asked if I'd sold it yet. I won't tell you what he said when I told him we hadn't even cut it! As we worked away in the dark up on my very top field with 360° views, I could see tractor lights working away in fields in all directions. I knew they were either harvesting potatoes or sowing next year's crops.

I've noticed a big change in public perception this year. I've heard commentators discussing inflation and citing 'the bad harvest this year' as an important factor. I think that people are actually reconnecting with the land and that's so fundamental to life, it can only be a good thing. I've said before that people think that food comes out of supermarkets but it doesn't, it comes out of fields. And if we get a lot of mud on the road while we are doing it, please bear with us.

**NOVEMBER 3RD 2012**

There was a farm sale locally last Saturday. One of those in the group that meet in the pub was reducing his farming operations. It's an interesting thing, a farm sale. All the minutiae of your farming life are laid out bare for all to see. Hundreds of people turn up to the sale. They poke about in every nook and cranny of the buildings. There are so many people there that if I were a thief I would imagine that every farmhouse and every farmyard for miles around would be deserted and wide open to a thief's activities. We all know how much good most farmers are at security!

So the sale day comes and goes but that's not the end of it. The post mortem in the pub goes on and on. It's almost worse than the post mortem after a game of dominoes. 'How much did the plough make?' 'I should have bought that.' I can't do with it, this examination of hindsight. If you should have bought the plough, well, you should have just gone on and bought it. 'How much did the combine make?' Here we go again, I've had enough of this. I interrupt. 'I saw _____'s car outside that house at the end of the village, it was there three hours.' I have their full attention. Ploughs may be interesting, but not as interesting as gossip. So I made it up: since when did that matter?

**NOVEMBER 10TH 2012**

There's no point, at the moment, in washing my car. There's no point in taking it down to my Eastern European friends. They probably miss me a bit but not that much because they give you a card that says you get a free wash after ten visits. I've had the card over two years. There's no point in washing the car because around here it's potato harvesting time. There's lots of wet fields and lots of potatoes and lots and lots of mud. Most of this mud is on the roads. I've even seen loaders shovelling mud off main trunk roads. I get the feeling that the potato harvesting is drawing to a close, with only the wettest fields still to do. But the end of the potato harvest won't signal a return to occasional car washing because we are just at the beginning of the maize harvest. That will bring even more mud out on the roads.

It's not a problem, potatoes or maize, for me or my car. It's all food and food has to be harvested.

There has been so much damage done to potato fields in this rain and so much damage still to be done to maize fields that it will not surprise me if the acreage of both crops is down next year. Some potato fields around here need the attentions of bulldozers to level out the fields. Most of these fields are due to be sown to winter wheat and I can't see that happening this year, which is in itself, an unwanted legacy of the wet weather.

What does concern me about the potato harvest is that there are some very large operators around here, people who grow perhaps 1,000 acres of potatoes on land that they rent just for one year (potatoes needing new land), and that have to go long distances to find this land.

Weather permitting, they harvest potatoes 24/7 and so it is a familiar sight to see huge tractors and equally huge trailers carting potatoes long distances day and night. The gross weight of the tractor, trailer and crop is often in excess of 30 tons and they can travel at speeds of up to 50kph. They are mostly driven by young boys who work very long hours, for several weeks. As we know, young boys are often on their mobile phones.

If I was one of those young boys I'd just love to do what they are doing. Big tractor, long hours, good money, but I fear that the combination of weight, speed and fatigue are recipes for a serious accident just waiting to happen. The sort of accident that could have repercussions for all of us in agriculture.

\*\*\*

There's this 14-year-old boy comes here one day a week from school on work experience. He mostly works with David, he'd never been on a farm before but he seems quite keen. But one day there's a job to do with me so we squeeze together in the loader tractor and go a couple of miles to clean up some spilt wheat. I find he asks a lot of questions which is fine with me. Eventually he asks if I like being a farmer. This somehow throws me a bit. At my age, the die seems to be cast, made my choice and I hope I've made the best of it. Also, at my age, there aren't many new career opportunities opening up in front of me. But he's due a reply. 'Yes, I love it.' 'Good,' he says, 'that's really cool.' And we drive on. It makes me feel quite good, described by a 14-year-old as cool.

\*\*\*

There are two pools (small ones) amongst the trees in a little valley on the land I rent. This year the Keeper put some mallard on there, I'm not sure how many, looked like a 100 to me, and it was nice to see them there. The next time I had a good look there seemed to be only about 20. So you ask, 'Where have they gone?' I do this with some trepidation because as I ask the question he puts on his glum face. He can do glum almost as well as I can do grumpy. 'Badger had them,' he says. 'I caught him at it. Took over 70 or 80.'

Now we all know that the Keeper knows more about wildlife than me so it was with much care that I asked, 'But how would a badger catch 70 or 80 ducks, presumably some of them would be swimming?' 'Easy, I saw him eating them.' But as he says it, there's a flicker of doubt across his face. I don't see him for a couple of weeks but when I do, it's 'I found out what was taking those ducks.' (No mention of having got it wrong.) 'I put one of those cameras up that takes a picture when something crosses its beam. It was an otter.'

This is really good news for me. I've never seen an otter in the wild and would love to do so. The news that there's one about on my land brings the possibility closer. I ask if it's still there but the Keeper thinks it's moved on. 'Boy,' he says 'you wouldn't believe what comes to drink at those pools, there's fallow deer, muntjac, lots of badgers, foxes, stuff that I didn't know was about.' It's quite interesting, no it's very interesting, that there is wildlife about that we don't see and that some of it exists, especially the deer, without the Keeper knowing, is amazing. I toy with the idea of buying one of these cameras. I have a farming friend who has one. Like me he is fascinated by wildlife and he puts it out all the time in different places, and like my Keeper, he sees on camera, lots of wildlife he didn't know existed on his farm.

Last time I saw him he was telling me that deer come on to his yard every night but he never saw a glimpse of them in the daytime. I asked him where he had his camera at the moment. 'It's on the diesel tank.' So you have to ask why. 'There's a meter on our diesel tank and there's 4½ litres going every night.' 'Why 4½ litres?' 'Well that's the size of an old-fashioned oil jug.' Must get one of those cameras – or perhaps I can borrow one.

## NOVEMBER 24TH 2012

This week's *Panorama* programme on culling badgers is difficult to ignore. If you had an open mind, and let's be honest, there are very few open minds on this issue, you would have watched the programme and have had difficulty reaching a conclusion. I thought it was important to see cattle actually being shot with a rifle because that is happening every day on farms. I hope that people found it distressing because the farmers involved find it distressing as well and in some ways it all comes down to shooting: it just depends which species is shot.

I have my own view and here it is. I think some badgers have TB. I know that badgers can spread TB with their excrement. I know farmers that keep their dairy cows housed all the year round because it's the only way they can keep badgers away from cows. Consumers think that cows should graze grass out in the fields. I know that TB is a disease of overcrowding and poor living conditions regardless of species. I know that if a badger has TB it will give it to all the other badgers in the sett because they all live in close proximity underground. So if you want badgers left alone, I want you, society, to go into my fields once a year, to use tranquilliser darts to restrain the badgers that are amongst my cattle, to test them for TB and if they have TB to cull them just the same as you do with my cattle. If cattle are mixing regularly with TB-infected badgers, there will never be an end game for the farmer because cattle will be re-infected forever. I will not pay for this testing of badgers, because they are not my badgers, they are yours.

And I have a word of caution, because emotions will run high between now and next year's proposed cull. Animal Rights issues have a history of attracting some nasty people. The chief executive of the RSPCA milked the applause at a public meeting by saying he would name and shame farmers involved in a cull, and he needs to be quite clear about what he is doing there. He tried to back-pedal later on in the programme by saying he would only name farms. How ridiculous! We've had people coming here for B&B who have been completely lost and asked someone ten miles away where our farm was. 'Oh, you want Roger Evans.' So much for anonymity!

I was at one time chairman of a large dairy cooperative. I think we had around 3,000 members: I'd met most of them, but I didn't know them all. I didn't know one of them also bred guinea pigs. Why would I? The drivers of milk tankers were threatened by association, they worked for a national haulage company and the directors of that company were identified and their young children harassed as they left school. My name, address and personal details went on their website. I've never been scared of anyone, well only my secretary, but it's not a nice feeling.

Milk collection was stopped immediately, the farmer had to milk his cows and throw the milk away until he could sell the cows. The people doing the threatening seemed to be beyond the law at one time, then they dug up the remains of one of the farmer's relatives in the churchyard and went a bit too far. We need to be mindful of these sort of people. They are driven primarily by a need to be nasty. Animal rights is just a convenient outlet for their nastiness.

\*\*\*

There is another issue with badgers. As we have established that they are not my badgers, they are your badgers, it is not my issue, it is your issue. Badgers seriously predate the nests of ground-nesting birds and kill leverets – and hedgehogs will be extinct in 20 years. Organisations like the RSPCA and the RSPB will not acknowledge this fact. They know it is true but they will not come clean on it because they don't want to alienate the people who will leave them money in their wills. Time will prove me right but my voice is presently in a minority. Anyway, what do I know about nature and wildlife?

\*\*\*

At long last we get a day that's dry enough to sow our winter wheat. It's just, only just, dry enough. The field has been ploughed a month now and it's my job to work it down and Stephen follows with the drill. Once I've got the power harrow sorted, right depth, right speed, right degree of tilth, I can settle back to an 'up and down the field' routine. The radio signal in this field is poor, but I'm not bothered, I'm ready for the wildlife.

There's usually plenty about when you're working a field, so bring it on. I've even got my notebook and pencil so I can make notes and share the wildlife with you. It's a 20 acre field and after two hours all I've seen is two crows!

Then out of nowhere a red kite drops down. I often say that being on a tractor gives you a privileged view of wildlife. The harrow I'm using is four metres wide and I pass the kite several times at close distance as I go up and down the field. What a beautiful bird! When they have located a good source of food there can be dozens about here and it's good to have this opportunity of a close up view. Then, next time up the field and it's gone. And there's nothing to report for another hour until out of the hedge at the bottom of the field comes a very fine white cockerel. It's one we asked the Keeper to put in his woods to keep the pheasants together. Stephen waves and points as his tractor passes. It's a bit of a local joke, these cockerels, we had 100 spare and offered one free to everyone we met for several weeks.

Next time down the field and there's three of them. A fine sight indeed, plumage blowing in the wind, combs and wattles bright red. Stephen thinks it's hilarious: the cockerels were put in adjacent woods and have obviously called to each other and joined forces. Next time down and there are six. This is an even finer sight but not quite as funny because their big feet are scratching up the wheat we have just sowed and they are eating it. It crosses my mind that these birds have seen a huge change in lifestyle. They were reared in an intensive shed with 20,000 other birds and now they are living wild in the woods and doing very well. It's a celibate lifestyle unless they can catch an unwary hen pheasant, but thriving they definitely are. Just hope that the Keeper doesn't bring them back to me when shooting is finished.

## December 1st 2012

Here, I go again, talking about television programmes, *Country File* last night. It didn't have my full attention because I was reading the Sunday papers, but I could pick up the theme. They were planting trees in Leicestershire that would form part of a National Forest. So that's fair enough, trees are our sort of lungs because of their vital role in converting carbon dioxide into oxygen. But there's always a bit that jars your subconscious and causes you to lower the newspaper and pay closer attention. 'In 30 years time these empty fields will be a growing, living forest.' True the fields were empty, but they were stubble fields so they hadn't been empty for long as they had clearly grown a cereal crop this year.

Yet the statement implies that the stubble fields had been doing nothing. And there's a bigger picture, a global picture, and like it or not, what goes on around the world is as important to us as what goes on in Leicestershire. If you accept that the world population is growing at an accelerating (some would say alarming) rate, and if you accept that all these new mouths have to be fed, then it seems very logical to me that if you plant trees on good arable land in the Midlands, then somewhere else in the world, someone else will cut down a similar area of rainforest to replace those arable acres, those 'empty fields'. So instead of feeling all cosy and warm about a national forest, we should actually be alarmed. And just as I lifted my eyes over the top of my newspaper to take in this story, the people who are organising tree planting and the people of *Country File* need to lift their eyes to a wider horizon, the global horizon, and ask themselves if they are not actually damaging that global environment. The environment needs to be linked to food security for the future.

***

There's a lad gets in the pub who has just been to an agricultural show in the Ukraine. He and his room-mate found phone numbers at the side of their beds. They thought they were dial-a-pizza numbers. They used them, two women turned up, but they weren't selling pizza.

### DECEMBER 8TH 2012

We cleaned out our loose housed shed the other day. Next day was one of those horrendous wet windy days we've been having lately. Good day, I decide, to cart the muck to our furthest field. It's so far it's a load-an-hour job, warm and dry in the cab and only five yards to go between tractor and loader, out in the weather. So I'm pulling out of our lane onto the main road and everything stops, not the engine, just the wheels won't go around.

It took half an hour of push-me, pull-you, to get the tractor back home during which time the road was completely blocked to lorries and vans. Cars could just get by. Lots of impatient people about! Turns out that because I'd got the lights and beacon on (foggy day) and I'd got the fan on (it all mists up otherwise) and (all right!) I'd got the radio on as well – the battery dips down half a volt or so, so the solenoid won't tell it to go, or something like that. I always tell you about my downs. For 18 months I've had this beautiful Jag that I bought on eBay. Cheap. So nice I was asked to do a wedding with it. I get in it the other day and the display on the dash says it's low. I knew this because when I got in it, it was squatting down on the floor like a hen squatting down for a cockerel to ride it. So it's got air suspension and there's a sensor on each corner that tells a computer how much pressure there is and the computer tells a little compressor where to send some air. Except that they've stopped talking to each other. So we make enquiries and a second-hand computer and sensors will cost over £2,000 which is without fitting and is about half of what the car cost anyway. And unless I'm mistaken, haven't Jags been going about with coil springs for years and years, quite happily anyway!

***

As you know, I like watching birds, and if I move on and ignore your comments about what sort of 'birds', the most numerous bird around here, by a large distance, is the pheasant. And they intrigue me. There's a bit of me thinks they are stupid birds: they have no traffic sense and are killed on the roads here on a regular basis. They are particularly vulnerable when they come onto the road looking for grit and especially when the acorns are dropping.

Yet there is another side to them, quite a canny side. I notice this in late afternoon as the light is starting to go. A time when a sensible pheasant is thinking about roosting for the night. I suppose the track through our rented land is nearly two miles long and as dusk approaches it becomes a thoroughfare for groups of pheasants making their way 'home'. There is a sort of wandering style to their movement – they are still looking into the grass verge for food, but if it's not a contradiction, there is also a sense of purpose in that they know quite clearly where they are going. I've often stopped the truck to watch them, because the light may be going, but the pheasants seem to know where they are going, and what I can't quite work out, is where? What surprises me is the distance they can travel in quite a short time. I know where all the woods and thickets are on this estate, but these pheasants will often pop over the boundary hedge and make their way to a small plantation half a mile away.

On shoot days it is a very different picture. The whole order of their day has been violently disturbed. Groups of birds are to be seen everywhere on the track; they are scurrying about now. Sometimes groups will pass each other going in opposite directions, and although they are obviously bewildered by the day's activities, they always stick to their distinct groups. Coveys of partridges are everywhere, they are always completely disorientated and you get the feeling that the partridges will wander around lost until it gets dark and then they will just squat down wherever they are.

***

I've got these calves, they are still outside on grass, though we will bring them inside this week. I give them some cake every day to keep them growing. They are in a big 25 acre field on the side of a quiet lane. There's probably a car along there every ten minutes or so. I've noticed that they completely ignore all other traffic but when I come around the corner they are already running towards the gate because they have heard me coming. This intrigues me because the old truck we use around the farm is very quiet, it's got a petrol engine converted to gas. It's a habit that stays with them for life because the next age groups up are a bunch of heifers that will go with the bull next week. They too are still out at grass but alongside a much busier B-road. I notice that when I go to see them, I can be at the back of a line of traffic, (which has all overtaken me), but the heifers ignore all the other traffic and continue grazing until I get within a couple of hundred yards, when, as one, they all lift their heads and start towards me.

Move on to the next age group, the in-calf heifers, and they are in buildings we rent. The men who work there reckon the heifers there can all be lying down contentedly, but they will all get up and start mooing, (we call it bawling) and then five minutes later, I will drive into the yard. It's nice to be appreciated. I think that's what it is.

## DECEMBER 15TH 2012

Vegans intrigue me. They promote their way of life with a sort of evangelical zeal. In the past I've been in negotiations with groups concerned with the welfare of farm animals. I'm concerned with the welfare of farm animals, I concern myself with it every day. But there is no negotiation to be had with vegans, because if you dig down into their philosophy of life, you often find that they are driven by their vegan agenda. There is no common ground because their ultimate goal is no farm animals at all. This position often muddies the water on planning applications for livestock units of all sorts, they will object come what may, regardless of any planning criteria, in an attempt to stop livestock farming on any farm, and certainly one that seeks to expand it. There's an advert for cat food currently on television that reminds its audience that cats are indeed carnivores, that they have teeth that were designed to eat meat. Not vegetarian cat food. I thought it quite a bold move.

***

It's getting towards late afternoon, it's been one of those awful wet windy days and there are several things that have gone wrong during the day that have put Stephen who works here in a grumpy sort of mood. A parcel van splashes its way up the yard, the driver sort of nods to Stephen as he passes, but Stephen calls out 'Oi, where do you think you're going?' The driver tells him that's he's going on down the lane to some isolated cottages.

Stephen tells him he can't get down the lane, he'll have to go back and around another way. 'But Tom Tom says it's only 300 yards down here.' And it's true, and it is a council lane, but it's about impassable now. The middle of the lane is so high, normal vehicles will bottom out, the hedges are overgrown on either side so you scratch your paintwork as you pass by. And there's a tree leaning across it. We used to go down in the truck without a problem: we were higher than the middle, the action of the hedge was as near as the truck got to being cleaned, we could just get under the tree, but at some time it had settled itself a bit more comfortably and we put a spectacular dent in the roof. Quite a talking point this dent – people often ask, 'How the hell did you do that?' So Stephen finds himself in this animated discussion with the van driver, who apparently believes that if Tom Tom says it's OK, then it is. Stephen: 'Well you go then but don't you come back here in half an hour expecting me to pull you out, you give Tom Tom a ring and ask him to pull you out.'

***

As I sit here writing this, the Christmas tree is grinning at me. It knows I hate it being there and it thinks it has won. My wife has spent the last two nights fiddling with the lights and she can't get them to work. I've taken the fuse out of the plug, it will stay safely in my pocket for another ten days yet. But it's not all bad news. Last night we had our charity Christmas bingo. I played. Never played bingo, it's quite challenging to start with. Then I won a 'house'. Life doesn't get much better than that.

### DECEMBER 22ND 2012

Today I'm going to try to draw the threads of two stories together but before I begin that process, I start with a digression. The digression starts at 4am in the morning when I leave the house to go to London. Leaving the house to go to London, or an airport or just to drive a long way, used to be a part of my daily life. In the summer it was just great to be out there on quiet roads in the early dawn, but it was always hard work in the winter when nearly all of your travelling was done in the dark. So today is one of those hard work days and even before I put the kettle on, I start the car to clear the frost away. Mustn't drink too much tea mind, it costs 30p to use the loo on London stations!

As I leave the yard, my son has all the lights on, and I can hear the tractor going as he clears the yards prior to milking. It's the same story a mile down the road where my neighbour, same age as my son, is doing exactly the same thing. Those are the only signs of life that I see for ten minutes, no other vehicles, nothing. They get up at this time every day. The people who fix the price of milk don't have to.

Anyway, to the story. I'm off to London to attend the ceremony of the Christmas cheeses. I'm a member of the Dairy Council and the Dairy Council organises the presentation of a large quantity and variety of UK cheeses to the Chelsea Pensioners at the Royal Hospital every year. I took my daughter and she thought it one of the best days out ever, one of those special days in your life that you always remember. You remember the splendour of the state apartments, the ceremony itself, but most of all you remember meeting the pensioners.

We had a guest of honour this year, a soldier who had lost both legs in Afghanistan a couple of years ago. He had had the courage and conviction to overcome those terrible injuries as best he could. He had even been pronounced dead in a field hospital but someone had seen a flicker of life as he lay on a stretcher and had revived him. Two years later he had competed in the Para Olympics with success. What a humbling experience. I've always tried to live my own life by the maxim: 'Don't let the bugger beat you.' I think I've lived up to it quite well but my modest achievements are as nothing at all compared to this brave soldier.

And I try to link that to something else. I have friends who have a daughter who is married to a serviceman. She has a role on the camp where they are based of supporting servicemen's wives and families. They showed me a piece that she had written in a magazine that is produced as a part of that support. There were two words that she used together a couple of times that stuck in my mind: 'your soldier.' 'When your soldier is on standby to go to Afghanistan.' Is there not a powerful eloquence to those words? There's an eloquence of ownership, a recognition that somehow we are all in this conflict together. So not only are they the wife's soldier, they are 'our soldiers' as well, all of them, just like the soldier at the cheese ceremony who has shown such courage and those soldiers who don't make it back as well, all of them 'our soldiers', and aren't we proud of them.

*** 

The hare coursers have stopped their activities, the main reason being, obviously, no hares, but a Keeper who phoned the police happened to mention that he thought the coursers had guns. This precipitated a visit from an armed response unit which was very prompt and very scary. Scary enough to frighten the lives out of the would-be coursers. Worth remembering that.

*** 

I've got six yearling heifers out on grass. I feed them cake every day to keep them growing and to help them through the wet days and the cold days. When I go to feed them I always talk to them: a 'good morning' or whatever is appropriate. Thus far, they haven't replied. In a field just the other side of the lane, a neighbour has had 13 dry cows and heifers and he feeds them every day as well. As the gates are opposite each other, his cows come down to the road when I am feeding mine, they not being as clever as my cattle and not able to distinguish between a Shogun and a Discovery. So I always say 'hello' to them as well.

One day there are only six there so I assume that he has taken seven home to calve. But he hasn't. One day our feeding visits coincide and he tells me he had eight reactors to his TB test the week before: they have all gone for slaughter, and seven of the eight were out of that field just across the lane. That field is bordered by extensive woods on three sides. Is it any wonder we farmers suspect wildlife? Next time you listen to or read someone pontificating about bovine TB and some theory or other they have, just spare a thought for us farmers and the sort of knife edge we live on.

## DECEMBER 29TH 2012

Most of the friends I mix with in the pub, go on a local shoot. Some shoot, some beat, but they all socialise together and there's as much talk of port and stilton at lunchtime and hip flasks during the day as there is of the actual shooting. It all seems to be taken seriously but with great fun. The stories that endure are those about someone falling in a ditch, someone getting stuck on a barbed wire fence or what someone's dog has done to someone else's dog. At the start of the season, they all took themselves off to a shooting shop to buy new clothes so they would all look the part. At the end of the shooting day, the pub lies between shoot and home and that is where they all end up.

I only know the next bit from hearsay, but they probably get in there by four o'clock and have a really good time until about seven. Then there is a general move home, 'Must go and get out of these clothes,' and 'See you later.' But when I turn up at the pub later that night, it's sometimes like stepping into a vacuum and most of my circle of friends don't make it back out. They are probably asleep in favourite armchairs and I have to wait until Sunday for all the stories. And if it's a good story I will probably hear it again on Tuesday and Thursday and it will probably become a better story with the passage of time, but there's no harm in a bit of embellishment and it's still a proper shoot, they have a mixed bag of around 50 and it's all shared out and it's all eaten but I suspect that it's the company and fellowship that is most enjoyed.

*** 

Our latest fund-raiser for our pub charity last night was a quiz. We've had an event every month since the summer and it's interesting how these different events bring out different participants. As a generality, most newcomers don't go to the pub at all. It's not that they don't drink. You should just hear the bottles rattling when the recycling van goes round. They must do all their drinking at home whereas I very, very rarely drink in the house but go to the pub regularly for the chat. But a quiz will get the newcomers out, they like to flex their intellectual superiority, because as everybody knows, newcomers are a lot brighter than those that are more local. A man at a local garage, having just solved an obscure problem with my car, once told me that he wasn't as dull as he looked. I told him it was a bloody good job. I like to think that I come into the same sort of category. So in putting a team together I just happen to ask my next-door neighbouring farmer, whose wife is a retired headmistress and who is as sharp as they come. She asks of me if she should bring her husband and I say, of course. She says that's good, as there might be questions on barbed wire or wheelbarrows. At half time, we are lying third, which causes some surprise, and a few say, 'Well done Roger' but what they really mean, 'He's just a farmer, he must have added the scores the wrong way up.' In the second half we play our joker and win the contest by some distance. They can't believe it. There's muttering and head shaking going on. It's completely thrown them and some of them will need counselling to come to terms with it.

## JANUARY 5TH 2013

As the year completes its cycle and draws to its inevitable close, the wheel of the farming year continues and next on the list for the farmers in the pub is scanning their sheep to see what sort of lamb crop is in the pipeline, or to be correct, in the uterus. One young lad in the village who keeps about 30 pedigree ewes has already done his and has 176% waiting to be born. But there's a sense of foreboding.

This year, along with the weather, farmers have to contend with the Schmallenberg virus. I'm not big on detail on this: it's a virus carried by insects and brought to us courtesy of favourable winds or, in this case, very unfavourable winds, from the continent. I'm not sure why it's called Schmallenberg. Probably the name of the person who discovered it. It's not a name that trips easily off the tongue and to be frank, it would be a lot handier for me to say it had been discovered by somebody called Smith. It causes deformity in foetuses and abortion in sheep and cattle. There's no treatment, just the hope that the animals will build up a natural immunity, but to build up a natural immunity, you have to be challenged by exposure to it. Thus far it has been mostly seen in cattle, but that's only because cows are calving all the time whereas sheep are only just starting to lamb in some early flocks. Where there have been problems with cows, there have been tests done for Schmallenberg and in every case I have heard of, it has been positive. This means that it has become much more widespread than first feared. The stories of where it has been found in early lambing flocks are horrendous, with lambs so deformed that they cannot be born naturally and the lamb is usually dead and the ewe dies as well. So, to return to the scanning results, there's not the usual point-scoring going on in the pub, just a sense of foreboding before their own results are known. Some flocks I have heard have only scanned at 50%, meaning half their ewes have aborted, with still the possibility of deformed lambs to come.

\*\*\*

The other big talking point has been Christmas dinner. A friend of mine always attends poultry auctions. Two years ago he phoned up to say, 'These turkeys are cheap, do you want me to buy you a couple?' In the end he bought me five for £25. It was generally felt that this was brilliant, but it wasn't, it was no good to anyone. In particular it was no good to the people who had reared the turkeys. Since then lots of people I know have attended poultry auctions in search of the elusive £5 turkey, but it's not to be found and this year they seem to be dearer than ever, and why wouldn't they be, with food cost up 40-50%? £5 turkeys devalues food and food shouldn't be devalued. If you remember, we had, earlier in the autumn, a surplus of cockerels about here. We sold them eventually for about 30 pence each, though towards the end it was 'buy a 30p cockerel and get another free.' Anyway, the first news of live poultry auctions brought the line, 'You slipped up selling those cheap cockerels.' Apparently someone who had bought 50 off us had sold them for £20. I was a bit sceptical about this, they were after all of a hybrid strain of poultry designed for egg production, not the table.

We have three of them roaming the yard. We wanted one of them to become another Neville, but compared with the legendary Neville, they are all wimps. They all roost over the calf pens and one dark evening I plucked one off his perch just to see what his condition was. They have plenty to eat, cow food, calf food, wheat, as much as they want. The cockerel I took off his perch must have thought his Christmas time had come but all I did was run my hand over his breast to see if there was any meat on him. There was none: his breast bone stuck out like a razor blade. If someone could fatten that up to make £20, good luck to them. So no regrets there.

But there's a more important issue: the turkey for Christmas dinner. For some reason I was determined to rear my own turkey for dinner this year. And I did. It's not been easy. I love turkeys, I love to see them roaming at will around the yard. The local foxes like to see them roaming about as well and to be honest they've had more than their share. I did end up with seven turkeys at Christmas, but only because I made a pen and shut them in for the last two months. So the fateful day comes and two were to be dinners and five were to go on and breed next year's dinners and feed the foxes and roam about the yard. Stephen who works here was to have one and we were to have the other.

It's not quite as simple as that. Stephen's wife is a bit of a perfectionist with food and I could tell she was a bit sceptical about our ability to provide her with a nice turkey. The first Christmas after Stephen came to work here, I gave them a beautiful turkey and she told everyone about it until someone told her I had only paid £5 for it and then she went quiet for a while. So Stephen and I approach the turkey pen and I insist that he has first pick of the turkeys. It's the little details like that, that are important in life. Life is often all about damage limitation. By that simple act I was in the clear, whatever the turkey turned out like, because Stephen chose it. As it turned out, their turkey was delicious. Ours was sort of OK. My wife tried cooking if a bit differently this year, something she had seen Jamie Oliver do on TV. I reared, feathered and dressed the turkey, I just wanted it roasting in the normal way. I don't want him interfering in mine. But there's no need for any of you to tell my wife that the turkey was only sort of OK.

Christmas dinners may be important but you have to think of others. It's a great comfort to me that that nice Von Trapp family get safely away every year.

### JANUARY 12TH 2013

I always try to write about people without identifying them by name, I certainly never try to write about anyone in a derogatory manner, but there are inevitably clues to identity. Today there are lots of very specific clues, but only if you live locally. We've always been blessed around here with great country characters, they've been one of the great pleasures in our lives and very slowly time is taking its toll on their numbers, and I can't quite see where the next generation of characters is coming from. I was quite close to one character, probably because I used to sit close to him in the pub. He's been gone several years now but in the last year before he died, he laid a hedge for me and one day we sat by the hedge drinking tea, and chose a sapling to leave to grow into a tree.

I always think of it as 'his' tree and who could wish for a better memorial? I remember being in the pub after his funeral and we were all a bit down because he had died a lot too young, and the next character I will move on to said to me, 'He wouldn't want us to be like this.' Within an hour we were all singing. Several of us ending up standing on a substantial oak bench built into a bay window, and it collapsed under our weight.

Sadly that next character also moved on last week and memories of him abound. He used to do weddings (take brides to church in a smart pony and trap, pulled by a smart little grey mare called Dolly). I happened on them one day in our local town. It was a Friday afternoon, as busy as it gets with children coming out of school, lots of shoppers and the like. He was taking three very old men for a ride in the trap. Two were in their late eighties, one well into his nineties. (Our character was well into his seventies but compared with his companions, relatively young). His idea of a ride in the trap is something known as a pub crawl. But there was no harm in it, except that Dolly had a foal, a foal of about 12 months of age, and naturally the foal had to follow its mother, and it did. No head collar, no leading rein, just trotting along amongst the traffic and the shoppers. Sometimes on the road, sometimes on the footpath. No harm in it at all but it could only happen around here. He made his living with a bit of farming and a bit of livestock haulage, which he did in an old Bedford lorry.

Every year he used to put a row of bales down each side of the lorry and take about 20 local men to some trotting races about 30 miles away. His lorry was his great pride and I suspect it will feature at his funeral in a small village and they are organising buses from the next village where there is more parking. Apparently some friends were tidying the lorry up for the funeral and were cleaning out some of what might be called 'coddle' from behind the seats: old coats, thermos flasks, sticks, there might have been some tachographs there but I doubt it. One of them finds a jar in there, puts his hand in it, and comes out with a handful of grey powder a bit like 'readybrek'.

So he pulls the jar out and discovers it's from the crematorium. Turns out our character had an old friend in the village who often used to travel around with him in his lorry. When he died he was cremated and they didn't know where to scatter his ashes. My character said 'Give them to me, I'll scatter them up on, _____ Hill, he always liked it up there.' And the ashes have been in the lorry ever since and had travelled miles. You can travel a lot of miles in ten years.

\*\*\*

My fundraising activities came to an end on the last Sunday of the year, with our annual pram race through the village. Not so many prams this year and not so many people watching, but we have been raising money since July this year, mostly from the same people, and that's all fair enough. I haven't collected all the money yet but we won't be far away from £4,500 which is a lot of money from a small community, especially as you inevitably ask the same people for support, again and again. My partner and I in the pram race (a pretty girl out of the village, and why not?), dressed up as the Queen and James Bond in the opening of the Olympics. I was the Queen, probably because I'm a similar shape to the Queen.

I'll never do it again. Hell it was cold in a thin pink dress. A lot of women came up to me in the pub afterwards and criticised me for not shaving my legs. A lot of men came up to me and criticised me for not wearing a bra and not having a bosom. It's all a bit scary, men dressing up as women. Some of them, not always the ones you think, get a lot of pleasure out of it and the first item of clothing they look for is the bra. I was very busy after the race collecting money and organising prizes but some of the male spectators had a few drinks and started to come on to me. We didn't get first prize for fancy dress (we won it last year and didn't expect to) but I thought we looked good in a contrived golden coach and with two toy corgis, bought at Sandringham Estate shop. I was very sensible this year, I didn't fall over the fireplace and cut my head when I got home. People say I was more fun last year.

### JANUARY 19TH 2013

My family and friends often tell me I'm a Victor Meldrew sort of character. I'm quite proud of that. Events occur that amaze me. Yesterday there was an item on the television news that was telling us how important it was for young children to have breakfast at the start of the day. You had to agree with what was said: it was important for a child to start the day well nourished, it helped the child to concentrate on its lessons, it was important for energy levels and well-being. So to make sure that all children have this essential start to the day, a local authority was providing a breakfast for all primary school children because so many were going without. No mention of parental responsibility in this! I bet all those 'breakfast-less' homes have plasma TVs and playstations. What sort of nation have we become? I was a child in the post-war years, I can remember the sacrifices my parents made to keep us as well fed as they could afford. What sort of parent turns a child out to go to school in the winter without a meal? 'I just can't believe it!'

***

A young local girl held a 21st party in the village hall last Saturday. It was fancy dress and the theme was what you had always wanted to be when you grew up. The party emptied the pub for the evening and for a couple of hours there were only myself and the licensees there. Which gave us opportunity to talk about all the others. Early on people popped in for a drink on their way so we could guess what they wanted to be. Stephen who works here came in in loud braces and a deerstalker hat and said he wanted to be a farmer, but he was dressed identically to our local electrician, and was quite taken aback when people said they didn't know what he wanted to be. I was invited to the party which was nice, but not really my age group. So this young pretty girl comes in dressed in a boiler suit but with very high heels sticking out the bottom. So I ask her what she wanted to be and she said she'd always wanted to be a farmer by day and a prostitute by night. And without any ado she unzips her overalls and underneath she's wearing a basque and a little skirt. (I only knew what a basque was because I've just looked in one of those catalogues on the kitchen table – honest). She asks me if I think she'd be any good at it! I tell her her heels are too high for farming but her outfit for her other occupation is excellent and she'd do very well. I don't actually think she would do very well as a prostitute, there are too many around here doing it for free.

***

There was a time when my wife would refuse to wear a seat belt. We had endless rows about it, so that if we were off out for the evening or the day, the event would be marred by the row we had had at the start of our journey. In the end I gave up arguing about it, it wasn't having any effect, so I adopted a sort of 'you get on with it' attitude. Shortly afterwards she started putting the seat belt on. The logic of this still eludes me, I feel there's a lesson there somewhere but I can't work out what it is.

Seat belts have never been a problem to me. But only up to a point. Because I drive off-road a lot and that sort of driving involves getting out to open gates, I have developed a habit of not wearing a seat belt as I drive between blocks of land. I suppose it's because it's only a mile anyway, most of it on a quiet narrow lane and only a small element of the journey on a B-road. But I was jolted back to reality yesterday. As I pulled out onto the B-road there was a police woman turning into the lane. I could see that she had spotted that I didn't have a belt on and in my mirrors I could see her executing a three point turn. She must have thought there were rich pickings to be had here. The truck is filthy and battered. Twice in the last week people have asked me if it is fit to be on the road. It is, but it doesn't look like it.

I put my foot down hard and to my surprise (and the truck's) we are soon up to 60mph. I meet Stephen, on his way home for his dinner and he is surprised as well, because I usually tootle along on this journey at about 20. As I take a slight bend I can see the police car setting off after me in hot pursuit. Fortune favours the brave, because as I get to our farm lane there is a feed lorry just about to come out, I just get past him onto the grass, so when the police lady gets to our lane, her vision is full of lorry and she speeds past. But I'm sure it's only a temporary victory, I bet she'll be looking out for me. I need to check the tread on the tyres, mend that broken indicator lens and even possibly put the hosepipe over it. It will be a great help if she likes dogs.

**JANUARY 26TH 2013**

The news story that 50% of the world's food is wasted is shocking, yet believable. Food, after all, is perishable, so there will always be an element of waste despite Man's best efforts. But the key word is waste, neglect of an invaluable resource. Sell-by dates are a big factor, some of them quite ridiculous and unjustifiable. I regularly eat food where the due date is long gone. It's become a part of the UK psyche, to err on the safe side, by some distance. It's a bit like some of the EC regulation thrown at us. Elsewhere they seek to find ways around it, here we enforce everything to the nth degree and then add a bit more on to be on the safe side. 'Buy one, get one free' offers are also to blame, because people think the bit that comes free doesn't matter. Imagine scenarios where you have bought two packs of cheese. You've eaten one but the other's gone a bit hard, probably because you didn't put it back in the zip pack package you bought it in, so you throw it away, but it doesn't matter does it, because it didn't cost anything anyway, did it?

One of the roles I have is a seat on the Dairy Council. We promote the benefits of dairy products within a balanced diet, in particular their importance in the diet of young girls, to mitigate against osteoporosis in their later lives. We also have to mitigate the damage that can be done by organisations that may seek to vilify dairying because of another agenda they may have. Salt is another issue that is on the receiving end and cheese manufacturers continue to offer cheese at lower salt levels. It's all to do with balance: why pick on cheese when some people reach for the salt cellar at every opportunity? I assume that someone somewhere adds up all the salt content of food, divides it into the population and come up with a per capita figure for salt consumption in grams, which they aspire to reduce. But things are not as bad as they think. If 'they' chuck half the food away, presumably they must be throwing half the salt away at the same time!

\*\*\*

There is something of the lemming in farmers. If something is expensive, they all want some. You wouldn't believe the times someone has said something like: 'Hell, the ewes are dear this autumn, I think I'll buy some next week.' It's the same if someone does something; they all have to do it. Remember the story of the mischievous farmer in West Wales who was getting a shed ready to bring his ewes in to lamb. He had bales of straw to move and some machinery. He had to put the fertiliser spreader onto a tractor to move it and told his son to drive around the field with it as if he was actually putting fertiliser on. It was a bit early for fertiliser, first week of February, but when they came out onto the yard after dinner, they could see three other fertiliser spreaders at work down the valley.

Cutting hedges reminds me of this. A couple of wet days in the early autumn, the combines still, and someone will get the hedge-cutter out. In no time at all, they are all at it. The reason I mention this is because we are doing our hedge-cutting now. I leave it until there are no berries to be seen. There may be a harvest in the fields but there is also a harvest of fruits, nuts and berries for birds and animals and I'm not going to destroy it until it's all harvested or eaten. There are downsides to cutting late, the wood has little or no sap in it, so it cuts 'harder' and doesn't look as neat and tidy. But like a bad haircut, it only lasts a couple of weeks and soon either it doesn't look half as bad, or you are getting used to it. As the wood is harder, it is more likely to cause punctures but that's a price to be paid for putting wildlife first.

When I finished playing rugby (which my joints tell me daily should have been ten years earlier), I took to doing some cycling, and hated having a thorn from a hedge cause a puncture. When the contractor was cutting my roadside hedges one year, I went out with a broom to clear the worst of the debris off the road. I did about 20 yards and gave it up. Although I was working behind the hedge-cutter and within his warning signs, it was quite clear I would be run over and killed: cars and lorries didn't slow down at all.

The bit about hedges that disappoints these days is the scheme where we are only allowed to cut the hedge once every three years. It's driven by organisations like the RSPB who obviously think it's better for wildlife. I'm on their side on hedges for wildlife, but I see no evidence that these hedges produce more food. They grow away on the two uncut years, the hedge becomes less dense and provides a lot less cover and shelter for wildlife and livestock. When they are eventually cut, they are an eyesore, battered and beaten because they are too 'loose,' looking as if they were chewed by a rat. I reckon a late cut like mine is a better solution than the every three years cut, but then I'm only a simple farmer, what do I know?

### FEBRUARY 2ND 2013

I know that shooting is not everyone's cup of tea but it's big business around here and provides a lot of casual employment for retired people, for example, who can go beating two or three times a week. It provides them with the sort of company they probably miss; it is excellent exercise; and on top of that, some extra income. Whilst we are in the grip of this spell of hard weather, shooting keeps a lot of wild birds alive. The woods and coverts are full of pheasant-feeding devices which are in turn full of wheat. No pheasant has to go far to look for food. Around every pheasant feeder are an abundance of wild birds. Most of them can eat wheat and they are taking full advantage. Much of the cover they can find in the woods is planted to enhance shooting or in game crops planted for the same reason, and this, for wild birds, is about as good as it gets. Likewise our farm buildings are full of wild birds who find shelter and food, especially amongst any straw that we use. I get a lot of pleasure from the bird table just outside our kitchen window. I buy them wild bird food but they get all the scraps the dogs won't eat (which to be honest isn't much) but when I mix any bacon fat with stale bread for example, the dogs can smell it and hang about the bird table all day.

\*\*\*

I've got 31 dry cows up at the top farm. During the day they graze a crop of kale next to the buildings. They go back into the buildings at night. Mid-afternoon I go up there and put fresh straw out for them to lie on and fresh silage for them to eat at night. When I go up to loose them out in the morning they come out eagerly to eat their kale, regardless of the weather. When they come inside at night, they are just as eager to go inside to their warm shed. It's working well – the cows obviously like the life they have. So my first job every morning is to go up there in the truck to let the cows out, just as soon as it starts to get light.

Yesterday was the first day of what has turned out to be a big snow. It had been snowing here a couple of hours when I set off and there were already three or four inches, but I was quite surprised when I turned off the main road onto a lane, to see that a snow plough had already been at work. Within half a mile I had caught it up. It's a snow blade mounted on a four wheel drive tractor. There are a few of these in our area where we have lots of narrow lanes, lots of hills, and the farmers keep these passable. They never get salted but they remain passable to essential traffic like mine in 4x4 trucks, just. I recognise the tractor. It belongs to a retired farmer in the next village and here he is out clearing roads and it's still not light and blowing a blizzard. I'm not sure exactly how old he is but he must be 85 or 86. I just think he's amazing. When I go out in the afternoon to let the cows back in, it's still blowing a blizzard, the snow is drifting across the lanes through the hedges and gateways. I meet the same man going the other way, it's ten hours since I saw him this morning. He gives me a cheery wave. Remarkable.

***

When I write these notes, I try not to be offensive. Some of my stories originally involved bad language and all sorts of innuendo, but I try quite hard not to pass this on, even if the story itself suffers as a consequence. I should add that some of the bad language is mine. Anyway, with my usual delicacy, I move onto the subject of cockerels. My search for a replacement for the famous Neville continues. About three months ago we had cockerels, wall-to-wall, but for various reasons we are now down to two. I'm still not sure that we have a new Neville yet, but one of them is showing promise. In theory they should be living a celibate life but theories only go so far. The one showing promise is only celibate if he can't catch the other cockerel. And that's OK, in the world of tolerance that we now live in. But this cockerel is a bird of huge appetite and a relationship with another cockerel is not enough. He has taken a liking for turkey. Not to eat it, like we do, but as part of the harem he aspires to. When I let the turkeys out every morning he's after them. There's one stag and three hens and the hens come out for their feed and the cockerel comes running. The stag fluffs up his feathers, huffs and puffs but is totally ignored by the cockerel whose attack is a bit like that of a big cat on a herd of grazing animals in Africa. In the pandemonium he causes, one turkey hen will break from the group and he will single her out to pursue, which he does, away from my eyes, behind the straw shed. And if I don't see what happens next, I don't have to tell you about it, so my high standards of delicacy and propriety are maintained.

## FEBRUARY 9TH 2013

It's not a day for hanging about. We're up at about 900 feet, the snow is about ten inches deep and there's a bitter wind blowing. Stephen and I have come up here to see to the dry cows, bed them down, put some silage bales out and fetch the cows in off the kale. The Keeper hasn't been hanging about either. Stephen draws my attention to where he's turned off the track through a gateway into a field, and spilled a fair bit of wheat onto the snow. The wheat is mobbed by small birds, we reckon there must be about 200. There's all sorts of finches there and I can see some yellowhammers. As if at some signal they all disappear into the adjacent hedgerow. It's not us that's scared them, because we are sitting in the truck and haven't moved for five minutes. Then I see a solitary red kite approaching. It's sailing on the wind, with hardly a movement of its wings. It is surely the wildlife equivalent of one of those drones they use to target terrorists. Swift, silent and deadly. It passes on up the valley and the small birds are soon busy at their feast again.

\*\*\*

Cock pheasants set up territories in the spring. Well, it's not spring, but we have this magnificent cock pheasant who has set himself up with a territory around the sheds at the top farm. He sees you coming up at a distance, and as you drive up in the truck you can see him running alongside you, just the other side of the hedge in the field. All of this story has taken place in the snow. As you busy yourself with your feeding and bedding down, he follows your every step and is never more than a yard away. If he has choices, if there are two of us up there, he prefers the company of Stephen and it's quite a picture to see Stephen going about his work with this pheasant following him closely like a pet dog. It's now got to the point where, if Stephen stops walking, he can bend down and stroke the pheasant.

Yesterday we had to move the electric fence on the kale. We had to move it a fair way because there's a public footpath goes diagonally across the field and you have to have the electric fence either one side of the path or the other. It was minus eight here yesterday morning, with a strong wind blowing. The kale is higher than the electric fence so in the circumstances the best way to knock the kale down is to drive a tractor through it so the fence doesn't short out. It was about 240 yards down the field and obviously the same back and the pheasant was at the front wheel of the tractor all the way there and all the way back.

I've got quite fond of this pheasant and am thinking he should have a name, Stephen is less fond because it tries to peck him when he gets back on his tractor. But there are still two more shoots scheduled here before the end of the season and I worry that he won't survive. So far he has obviously kept his head well down. On Saturday morning I happen to see the shoot organiser on his way to the shoot. He likes to call himself 'shoot captain', but we won't dwell on that. After the usual pleasantries I tell him it's a big day for him. He wants to know why, so I tell him the story of the cock pheasant very much as I have just told it to you. 'If that pheasant is shot,' I tell him, 'there will be no game cover here for you next year, no acre of maize for your best drive, in fact no co-operation from me whatsoever.' His face is a picture. 'But what are we to do?' he asks, 'How will we tell which is your pheasant amongst all the others as they fly over the guns?' 'That's something you'll have to sort out yourselves, I've told you what the consequences will be.' And I drive on. Not big on humour, your shoot captain on shoot day.

That was Saturday, today it is Monday morning, I've not seen the pheasant for two days now, the inevitable seems to have happened. But late on Monday afternoon when I get the cows in for the night, he's back. It's the same pheasant but his behaviour is just a bit different. He still follows me about, but he's more cautious, he's obviously had a negative shooting experience. I wonder where he's been and assume he's been put over guns and ended up in a wood some distance away, which isn't difficult around here, and it's taken him two days to get back. By his cautious behaviour, he might have been sprinkled with a pellet or two. The shoot have decided there's plenty of pheasants left and they've put in an extra day on Wednesday and then there's beaters' day, which will involve about 20 beaters armed to the teeth, resembling a remake of *How the West was Won*. My pheasant has to survive all that yet. If he does, I'm going to call him Norman.

## FEBRUARY 16TH 2013

I had one of those nasty colds in December, it seemed to last three weeks. No one goes for a haircut when they've got a cold do they? Then it was busy pre-Christmas and I couldn't be arsed with it. Who wants to be asked how much time they've got off at Christmas and if they're going anywhere nice for their holidays when they are busy? Though to be fair, the girl who cuts my hair isn't big on that sort of thing. If there's no one else in the salon

I get a haircut, a cup of coffee and more gossip than you would ever believe. All for £4.95! I usually give her a fiver and let her keep the change, known for my generosity, me! January has been much too cold for a haircut and I've reached the stage when people have started to ask when I will cut it. My doctor didn't cut his hair for 12 months because he was fighting to keep our local hospital open, (which he did) and he said he wouldn't cut his hair until he won. He went around for 12 months looking like a Herdwick sheep. It's called a community hospital now, before that it was called a cottage hospital. It's a wonderful place for people to recover amongst people they know. It isn't very popular with very old people because before it was called a cottage hospital it was called a workhouse. If it was a workhouse today it would probably be full of dairy farmers.

Anyway here I am with this long hair and I wonder if there's some grand gesture I can make for its removal. Not that I have to remove it, I could grow a pony tail, but I don't dwell on that. There's a guy comes to the pub occasionally who has a six inch long pony tail growing from the point of his chin. He seems to spend a lot of time trying to keep it out of his beer. To start with I said I wouldn't cut it until we had seven consecutive dry days. But that could be never, so I've amended that to when the cows go out, which could be considerably sooner. (February last year). It does look a mess and I've raided other people's shampoo supplies. I've even sneaked some 'leave-in' conditioner. Wonder what that is? Before my daughter married I had multiple choices every time I went into the shower. I could curl it, take out any frizz or bring out the highlights. Yesterday someone asked me if I'd always wanted to be a hippy.

**FEBRUARY 23RD 2013**

We have to endure all manner of inspection and regulation on our farm and I hate it. It is done in the name of Farm Assurance and was largely a consequence of BSE. BSE was a terrible prospect and inspection and regulation was a price we had to pay. Trouble is, BSE has gone away but all the regulation has stayed with us. There are lots and lots of things we have to do to pass Farm Assurance but during inspection there are always things we are asked if we do. If we say no the reply is, 'You don't have to do it but you really should.' Next time you have an inspection some of the 'shoulds' have become 'musts'. And so it goes on remorselessly getting more onerous. When we query all this, it is justified by saying it is all driven by the supermarkets on behalf of their customers. In the light of the horse-meat scandal, why, pray, should we allow the supermarkets to drive anything? Most of us wouldn't select horsemeat to eat by choice on a menu but it seems that inadvertently we have. How can they say it is safe?

Countless small rural abattoirs have been put out of business by just the sort of regulations and inspection that I refer to, such as having a vet present at all times. If they didn't know all these horses were being slaughtered how do they know how they were being slaughtered, where they come from and the conditions they were in? The answer is, they didn't, so how can they possibly say it is safe? They can't. Someone once said to me that you shouldn't eat meat unless you recognise the cut of the joint. 'Once it is all minced up,' he said, 'it could be almost anything.' Turns out he was right. I've always said that there are probably more rogues in the meat trade than there are in the second-hand car trade. Turns out I was right as well.

Off-hand I can't remember anyone getting into trouble over the BSE débâcle. Same old story, the rogues are smarter and more cunning than those who would police them.

*** 

We've had friends from Birmingham staying for a few days. One day they went off for a ride round and when they returned they were quite excited. They reckoned they had had an adventure. As they were driving home they came across a newish-type Mercedes car stopped on the side of the road, its bonnet was raised and the driver was waving frantically for them to stop. They stopped and they gave me a fulsome description of the conversation that took place. I won't try to rewrite the conversation because I obviously didn't witness it. The driver was hysterical: he told them he hardly had any petrol left and his credit card wouldn't work. He had a baby in the car (which he did), the baby was not well and he needed to get it back to its mother in Manchester. He said he had been in several of these garages that needed a credit card in at the pump.

I was only half listening to all this as I was reading the paper, but that was the first bit that didn't ring true, because around here I don't know of any garages like that. Anyway the man was almost down on his knees now, begging for some cash, for which he was prepared to give them the ring on his finger. He took the heavy ring off and showed them the 18 carat mark. My friends, suitably sympathetic, gave him £20 and took the ring in return. The man said that if they had any more money he would give them the gold chain around his neck. They declined this offer but gave him their address and said that if he returned the £20 they would return the ring.

To cut a long story short, as it's turning out to be a long story, my friend used to work in jewellery in Birmingham and reckoned the ring could be worth between £50 and £70. As soon as she gets home she takes the ring to a friend in the trade and starts to tell the story. The friend stops her story and carries on with it almost word-for-word. Turns out this was about the tenth ring this friend had been asked to value in similar circumstances. 'He must have been very disappointed with you, most people took the necklace as well and parted with £50.' The value of the ring? Perhaps a couple of quid, or you could always wear it yourself. Me, I'm trying to work out how many times a day you needed to work the scam to pay for a new Mercedes.

\*\*\*

I'm quite proud and fond of my turkeys, I have a stag and three hens left and they roam the yard wherever they wish. For weeks now we have shut them up at night but I didn't want to have to do that. I wanted them to find a warm dry safe place of their own choosing to roost. I didn't want to have to find them every evening and drive them to their pen. They don't drive very well; you never know where they would be and if you left it a bit late, being black, they were difficult to find. Enough was enough. So they were left to their own devices.

Came home from the pub one night and they were perched on a fence on the yard. It was blowing a gale, the rain was lashing down and it was bitterly cold. I couldn't believe that they'd chosen to spend their night in the open when there were so many warm dry buildings available. They sat huddled together, cold and wet, like so many vultures at the end of a bad day. But they were elsewhere the next night. Our next-door neighbour has a motor boat, not a very big one, but very smart. (It's got living accommodation!). They spent the next night perched on the roof of the boat. He wasn't best pleased. I told him they were not as bad as seagulls. Won't tell you what he said. Anyway the boat hasn't got its bottom wet for two years that I know of. I thought it was good to see it used.

### MARCH 2ND 2013

Just across the lane from some of my fields, my neighbour runs his dry cows and heifers. At one time, just over a couple of months ago, there were 13 there, all Jerseys. At his annual TB test, seven of these failed and were slaughtered. He took two home because they were due to calve. That left four, and those four have been there all winter. He hasn't taken them home, for obvious reasons, and now their time to calve is approaching. There is one heifer there that catches my eye: she has a lot more white on her than your usual Jersey and as she approaches her first calving, she is a picture. Fit and well, with a lovely udder.

Last week they were tested for TB again. They did it in a pen in the field, and two more failed, including the one with the white patches. I pass by there several times every day and every time I expect to see her with a calf. I don't know if she will ever calve because she is waiting to be put down, along with the other one. Adjectives fail me, it's such a tragedy, due to calve at any time and due to die for no fault of her own. Such a waste! What sort of society are we that allows this to go on and on? Every day my neighbour takes them hay and corn. Every day he stands there whilst they eat the corn and watches them. And I wonder what is going through his mind. I think I've got a fair idea. The field these heifers are in slopes up and away from the road, it's surrounded by woods on three sides, and if this were a cowboy film, it would be a box canyon. He's waiting for them to finish their corn because when they are finished he puts the troughs on top of the hedge to stop the badgers peeing in them.

When I went past this morning there were only two heifers there. They will have put the two down in the field. They were so close to calving, the calves inside would have been having a bit of a stretch and thinking of making a move. Just glad I didn't see it. Even more glad I didn't have to take part in it.

*** 

It's a dry sunny day and there's a bitterly cold wind blowing. I've just cut the lawns. It's the first time I have ever cut them so early in the year. It's not the first time I've wanted to cut them in February and it's not the first time they've needed cutting in February, it's just the first time I've managed to start the lawnmower in February. Usually it won't start so I take it to the dealer where they put a label on it and it joins the long line of other lawnmowers that won't start either. Putting a label on it is dealer-speak for: 'Don't expect it to be ready for at least a month'. But it fired up first time yesterday which is in itself I hope an omen for the year to come. The grass had grown a lot during the wet winter so it's good to get that growth under control.

Firstly I get number two grandson to rake out the mole hills of which there are a goodly number. He's not best pleased as it takes him an hour, which is an hour he would rather spend with my laptop. So by the afternoon, away we go. I cut the worst lawn first, quite a high cut and nice and slowly, the rotors spreading soil and grass in quite equal quantities. Half way through I stop to answer my mobile and I can smell the fresh-cut grass as it dries. When I finish I drive the mower back up the yard and put it in the shed. All I can hear now are the milking cows bellowing and roaming about their yard. They can smell the grass as well and it's unsettled them. We'll need more than a week dry before they can go out.

***

As I was in a mood for cutting and tidying up, I then turned my attention to my hair. I had vowed not to cut it until we had had seven consecutive dry days. We'd had five and the forecast was for two more, so I thought that would do. I phoned my hairdresser. First off, she said, 'I saw you walking down the street last week and wondered to myself what I'd done to upset you?' She'd never seen my hair so long (and neither had I). For years she has been trying to talk me into having what she calls a number 2, but I tell her it's still cold outside, there's still plenty of time left in this winter for some 'weather', so I'll just have a bit of a tidy up and leave it at that. You can tell how long it was, she gets two of those long grip things and fastens some of it on the top so she can get at it. Twice in one month now I've been called a hippy.

***

I was tidying up our work bench. Well, I wasn't really, I was looking for a particular spanner, when I came across our cow bell. It's a proper Swiss cow bell and was sent to me by a *West Country Life* reader. Last time I hung it on a cow, the cow was dry and grazing some fields we rent. I used to like to hear the bell ringing as the cow went about her daily life, it has a melodious sound. And it rang even when she was chewing her cud. I never gave a second thought to the cottages that back onto the fields. One evening an occupant phoned to complain. I thought he was joking and laughed. Big mistake! He got quite nasty and threatened to shoot the cow and take the bell down to the tip. I get a mischievous delight from this sort of thing in relation to my animals. 'Does that dog bite?' 'You bet he does.' 'Is that cockerel nasty?' 'No, he's a pussy cat.' 'Your turkeys have just chased me, pecked the putty out of my windows, done lots of poo on my boat.' 'That's what turkeys do.' Time, I think, to get the cow bell out and working. It's spring soon; alpine pastures beckon.

## MARCH 9TH 2013

My stag died last week. It's not a good time of year to lose a stag turkey. It will be spring soon and a turkey's thoughts will be turning to love and fertile eggs. My thoughts are already turning to the source of my next Christmas dinner. So my thinking is of finding a new stag, and fairly quickly. It's a sad sign of the times we live in, that my first reaction is to put my laptop on the table and google 'turkeys for sale'. And then I pause. It's a bit too easy. Most people would do exactly the same thing and search on the internet. But where's the fun in that? As a consequence, my thoughts turn to the hills. You can find almost anything you want in the hills around here, it's where all the characters live, it's where the stories are.

We were out with friends the other night and I was telling them about my turkey dilemma. Our friend, a retired school teacher, doesn't hesitate: 'So and so's have always got lots of turkeys.' There you are, up in the hills, told you so. But she's got another story to tell. 'I taught their son at school, couldn't do much with him, always a free spirit, always looking out of the windows across the fields.' And she goes on to tell us that his parents couldn't do much with him either.

As soon as he could crawl, he was away. By the time he started walking, they couldn't turn their backs on him for a minute, so they made a sort of harness for him that he couldn't get off and tied it to the washing line that stretched across a piece of rough grass that they called a lawn. 'By the time he went to school he had worn a pathway in the grass down to the soil so there was this muddy track underneath the clothes line. Wouldn't be allowed to do it today.' She reflects on this then adds, 'But it didn't do him any harm.' She says this wistfully, as if yearning for a bygone age when children had a place in life, even if that place was tied to a washing line.

I know all about hyperactive children, having had one of my own. My two children were at the opposite ends of the hyper scale. You could put my son to bed as a baby and he would sleep 12 hours. My daughter only seemed to need three or four hours sleep and she assumed that everyone else was the same. It became quite a big family issue, dividing up her care. As my wife had to try and pacify her all day, we slipped into a routine where I took care of her when she woke up in the morning. This was usually between 3-4am. I had an excellent lad working for me at the time and he had to do the morning milking to fit all this in. From a time when she could barely crawl I would go to get her out of her cot, a cot that would be bouncing up and down from the activity within. I would get a big broad smile and we would be off down to the kitchen, I'd dress her and she'd wreck the kitchen while I listened to the World Service. They were long hours, so long that a highlight for me would be *Farming Today* at a quarter to six, which is quite sad when I reflect on it. In the dark days of winter you just had to stick it out, but as the mornings got lighter I wanted, and needed, to be outside doing my work.

A pram or pushchair wasn't very handy about the yard: wheels would get muddy, blankets would be thrown off into the 'you know what', then one day I solved the problem. I put her in a bucket. We had some tallish buckets that had held dairy chemicals and a young child fitted in quite nicely. They were narrow, so she couldn't move around in it, as in topple over, and she seemed to sit in it very comfortably, sort of squatting in on her haunches. The top of the bucket came level with her nose so she fitted in it really well, her little hands on the rim of the bucket and her eyes peering over the top. In those days I had sheep and pigs to feed as well as calves so as soon as it was light I would be out there, a bucket of feed in the crook of one arm and a bucket of baby in the other. And she loved it. She was transformed, she just sat there, transfixed with what was going on. Watching me as I tended to my animals. I soon discovered that it was even better than I thought. Everywhere I went, into pens or into loose boxes, there were gates and doors. Gates and doors mostly have gate hooks so you could quite easily hang the child bucket on a gate or hinge where she would be perfectly safe whilst you cleaned out a sow and pigs or bottle fed some lambs. Looking back it was a time of bonding, father and daughter. We have always been close and some of that was probably down to a bucket.

Looking back, I think all little girls should have a bucket to put them in, to keep them under control. And that bucket should stay with them throughout their lives. As they grew, you'd just get a bigger bucket. As the girl becomes a woman they should still have a bucket. So when you needed those periods of peace and quiet, you could put them in there and you could get them out again when you were ready. Simple and so effective.

Can't think how you could improve on it, unless of course, the bucket had a lid.

### March 16th 2013

The roads around here are littered with the corpses of cock pheasants that have been run over. It's the time of year when they create territories that they defend. Because the east wind has dried the land up I can get on with carting muck and there are five dead cock pheasants on the road on my two mile journey. In a few weeks time it will be hen pheasants that take a turn at joining this carnage as they stagger off their nests in search of food and water.

For the cock pheasants it's a very attack-driven form of defence, which sounds like a contradiction, but it works like this. As soon as you come into their range of vision, they race towards you, and they do race, just as fast as they can run. While I'm on the tractor they attack the front wheels. They will often run 50-100 yards alongside the front wheel, so close you cannot believe it. Sometimes they run in front of the wheel itself, so close that you can't even see them, in fact you often think they've gone, then you get a glimpse, and they are still there. On one part of my journey, the pheasant is running flat out parallel to me but he's inside the field and I'm on the road: I'm probably doing 15mph here but he keeps up. Then the road forks and I go up the road to the right, so he's through the hedge, across a lane and through the other hedge to try to get at me. The cock pheasant that was living up at the buildings, I can't remember if I called him Norman or Stanley, has not been quite as bold recently, I suspect he's had a bad experience on the last shooting day. His territory is, or was, around our top buildings. He would do the same sort of chase each way as I took every load of muck. He would disappear from sight as he ran in front of the front wheel. If I slowed down every time this happened, I would never get any work done, but I try to keep track of him. Now I see him, now I don't. And now I haven't seen him for a while. I think he's gone now and given up the chase but I look in the mirror and there he is fluttering up and down in what looks like his death throes. He must have been clipped by a trailer wheel. I'm hoping he has just had a knock and will recover but when I take my next load he's dead alongside the track. There are two buzzards on a fence nearby, licking their lips.

***

It's 7.30am, it's bitterly cold with a strong wind blowing from the east. Undeterred, the dog and I are off in the truck; we are on important business. We are off up into the hills. If it was cold on our yard, now this is *really* cold. A man is waiting for us by some buildings; he has his back to the wind; the trees have their backs to the wind; even the buildings have their backs to the wind. Usual pleasantries exchanged, he indicates a poultry crate on the floor nearby. 'There he is.' It's what we are here for. A new stag turkey. We put the crate in the truck, it's no day for much hanging about, but he indicates the rest of his turkeys, over a fence, having their breakfast. There's a brown stag there, the brown of a spaniel and I've never seen one like it before. I ask where he got it and make a mental note to go and see if there are any left. My new stag had three days shut in to settle and he's now, as my grandson's put it, loved up. The turkeys are out on the lawn now and they do their lovemaking there every morning. He's a very considerate lover, he puts his arm around them and gives them a bit of a cuddle first. The actual performance seems to take five minutes and is closely monitored by his other wives as they wait their turn. You can monitor their progress by the bare patches they cleared in the frost on the lawn.

\*\*\*

I don't go down to the rugby club that much these days. At one time it was my life but now I go perhaps four or five times a year if there's a sponsors lunch. After I stopped playing I used to go to watch my son play and he is now doing the same watching his son, so I mostly stay at home and do some work. I was down last Saturday for a lunch, a good friend of mine is President and I was on his table. This is not the social honour it appears, it's just that one of his guests cried off and he was one short. So I'm a bit early and I get in a corner and watch what's going on. I can't get my head around all this kissing. It's become something of a norm in recent years but I'm still not sure of the protocol of it all. Most people come in, greet their friends, touch cheeks with the ladies and mime kisses, going 'Mmm wa.' And people, the other people, on the President's table, are much higher up the social scale, a different kettle of intellectual fish altogether, they do two cheeks. When I used to do business in France many years ago, it was always three cheeks. It's all very confusing. It leaves me in a dilemma what to do. So I go straight for the lips. Lips is nice. As they say in *Under Milk Wood*, 'Lips is a penny.' So why not try to get a free penny-worth?

## MARCH 23RD 2013

You can look over the gate as much as you like, but the first real look at your fields after the inactivity of winter, is when you start work on your tractor. So Stephen has had the first look around as he has put the first dressing of fertiliser on the grass fields. We can tell how late the season has become because the cows were already out at grass at the same time last year. Stephen comes back after doing the grass here at home and reports that he has seen two hares and three deer. The hare news is really good news. I might have hares wall-to-wall on my rented ground but for many years they disappeared here at home. For three years now I have seen hares here and can only assume that they are making a tentative return. There's a good chance that these two will breed so I am pleased by Stephen's news.

I've yet to see deer here personally. There's a lot of luck involved; they can skip away over the fences back into the wood long before you get near. But as you know, I like to see wildlife about. The same day, I'm watching the lunchtime news with my eyes closed and there's an item about deer numbers and the need for a cull. I wasn't paying enough attention to work out if they needed to cull a million or half a million. So who counts them and who are they kidding? How could you count them and how do you know you haven't counted the same deer twice? Stephen's just seen three, they went into the wood. If I go down there, there will be none. So that's an average of one and a half. Exactly. If they do a cull, I hope they leave my three alone.

\*\*\*

People are strange. Slow them down on their journey with a slow-moving tractor and you get everything from horn blowing to angry gestures. There are a lot of timber lorries that come past here and the drivers are aggressively impatient. They will overtake where they shouldn't, then cut you up so they can get back onto their side of the road. They leave you in no doubt that they would like to get out and fight you, if they only had time. If it's livestock you are moving, it's a bit the same.

I've had people wind windows down and tell me that I have no right to move livestock on a public road. When I tell them that people have been moving livestock on this road for hundreds of years, my sarcasm is lost on them. Some are so impatient, as they try to force their way through, that the dog is in real danger of being run over. They obviously don't realise how serious the repercussions would be if they did. If you are moving cattle, woe betide you should someone get some poo on their car.

Years ago we were moving quite a lot of ewes plus their lambs on the road on a Sunday morning. How were we to know that there was a cycle race on the road that morning? There was certainly no way we could know that one rider had made a lone break from the pack, of about half a mile. I won't tell you what he called us when he found his road blocked with sheep, and in no time, the rest of the cyclists had caught him up!

But our adventures on the road have taken a new direction recently. On an almost daily basis, our turkeys go walkabout. The furthest they have been is half a mile on a B-road. There is no alternative but to walk them back. It must be a strange sight for drivers, a man, a dog, and four adult turkeys walking along the road. But there's no impatience now. The drivers are all smiles, windows go down, 'Save one for me,' 'It's not Christmas yet, is it?' The stag fluffs up his feathers and stands his ground as each vehicle comes along, which precipitates more smiles. People are strange.

***

There is an unusual seasonal lull in the pub conversations. In the past we would have been flat out on lambing stories but almost every farmer who has sheep has delayed his lambing date. They have discovered, just like all livestock farmers, that animal feed is so expensive that it's better to use grass and forage instead. So if your ewes lamb when the spring grass turns up, they will milk well and lambs will thrive better on nature's cheapest natural food – grass. To fill this conversational vacuum, talk has turned, believe it or not, to moles. I think I started it off with my stories about the moles on my lawn. There's a couple of local lads go around on pest control: they do the rat baiting for our chicken sheds. They charge about £5 a mole and reckon they have over 300 traps down at any one time. £5 seemed a lot of money until I realised I could pay it on our chicken shed account. My son doesn't know about this so don't tell him. They caught the main culprit fairly quickly but immediately all my colleagues become mole experts. And the conversations soon go onto the subtleties involved. A lot of this centres on the need to avoid human scent on the actual trap. Some advocate the use of gloves although there's always someone who catches moles without gloves and with traps newly out of the shop. The consensus, however, is to use gloves and put the trap in a muddy puddle for 24 hours.

There is always a competitive edge to these conversations. Who can catch the most moles? But then the conversation takes an unexpected turn. Who knows the best mole catcher? We are told of a character in a village a few miles away. A catcher *par excellence*! He doesn't worry about human smell getting onto the traps as he handles them. It's quite simple really. He always carries a dead mole in each of the two pockets of his waistcoat. And for good measure he has two more dead moles in his jacket pockets. So he doesn't have to worry about his human smell because he already smells like a mole anyway. Even if it is a dead one. This is a show-stopper and it goes quiet as they all take reflective pulls on their pints. What they are thinking is: what does a mole smell like? And how often does he replace them with fresh dead moles? So many questions.

### MARCH 30TH 2013

My dog Mert is an important part of my life, he's a constant, always there, well, nearly always. We've been juggling about with our tractors, we've sold one and seek to part-exchange the very oldest for something a bit better. To put some context on this, the oldest tractor is 26 years old. That's not its only problem. I've been using this tractor a lot lately and there's no room on it for a dog. So I've been carting muck one way and fetching silage home for the cows the other, (different trailers), and there's no room to take the dog. I'm not sure he fully understands this. He knows what I'm doing and sits on the yard all day, rain or shine. As I approach he smiles at me and gives his tail a tentative wag, there's an imploring look on his face. As I drive past, without stopping, his look and demeanour are pitiful to see. I feel a complete bastard and he clearly agrees with me. But for Mert, all is not lost. We have a prospective purchase here to try, a replacement tractor, and I've put it on the silage trailer. Although it's 13-years-old, it's got low hours, is in apparently perfect order, you can see the outside is perfect, so mechanically you have to cross your fingers and hope for the best. But I work a bit on the assumption that if someone looks after the outside properly, there's a fair chance they changed the oil on time and serviced it in orderly fashion.

These are all issues you take into consideration when making a purchase, but very high on my own list is the fact that there is probably more room for a dog on the floor of the cab of this tractor than any other we have. I intend to buy this tractor though I will drag the deal out a bit longer yet. Then Mert can take his rightful place at my side – well, at my feet, anyway.

***

I'm in a 25-acre field of stubble that was in spring wheat last year. I have to leave 20 acres of it as fallow all summer for the birds but I can plough five acres and put in spring barley. So I'm up here in the truck to mark out a five acre block to be ploughed. It's all fairly simple stuff really, I just drive the length of the field and note the length on the truck mileage and do the sums, but woe betide you if you get it wrong. If there are spot-checks done and mistakes have been made, fines will follow that would really rock the financial boat. As the boat is already taking in water I can't afford for this to happen.

I drive the length of the field, it's a long one, check the distance and then drive back to double check. The two distances I now have are some way apart so I do it all again. I now have four different distances to work with and I've got to choose one and divide it into 5x4480 which is the number of square yards in an acre, if my memory serves me correct. But why would my memory be any more accurate than my truck? There must be an easier way. The field is a long rectangle, the same width all the way down, so I count the fence posts (which are two yards apart) and divide up, and the job's done. I put a marker on the appropriate fence posts for Stephen when he comes with the plough. As I reflect on my success I note six hares chasing each other about in the stubble, doing what hares in March do. They stop occasionally to watch me and then continue. I sit and watch them for quite a while. Why not?

## APRIL 6TH 2013

I told you that there was a bit of a vacuum in the pub conversations because most of the sheep farmers had delayed lambing until mid March. Now there is another, more serious vacuum to fill. The last few days here have been so distressing for the sheep farmers that most of them avoid going to the pub at all in order to avoid talking about it. This late March snow has caught everyone out. It's about eight inches on our yard: go to the farm we rent and it's over your wellies. What chance a week or ten-day-old lamb in all that? Very little. The ewes themselves are not in best of condition. They've had six months of wet weather and they don't thrive in the wet.

A young lamb's only chance of survival is a belly full of warm milk. If your mother hasn't got much milk and her udder is down in the snow, the results are inevitable. I really feel for these farmers, as I do for their sheep. Most of them lamb their ewes indoors and spend long hours of the day and night caring for them. But it's a sort of conveyor belt system that involves moving ewes and lambs out into the fields to make room for the next batch. Just imagine, if you can, the heart-wrench of trying to go out and feed these ewes and lambs, lambs that you have probably stayed up all night for, to bring safely into the world, and all you are doing is picking up dead lambs that have succumbed to hypothermia.

This is a big sheep area and hundreds, possibly thousands of lambs will have died here over the last week. No one will ever know how many because, as I've said, no one wants to talk about it. It's obviously bad for the sheep but just how demoralising will it be for those who try to care for them? Many spend long hours on their own, they don't want to tell anyone the extent of their losses so they are bottling it all up inside. It's tough, I just hope the mental scars will heal.

\*\*\*

The Keeper tells me that he is driving along the road through the estate and in the distance he can see some men and dogs hare coursing. So he drives to our town/village, and goes straight to the police station. There so happens to be a policeman on the premises, (this is an opportunity for sarcasm but I move on), and he reports what he has seen. The policeman isn't interested, he says it would be a waste of tax-payers' money. He would probably have to chase them, he would probably only catch them if he had their vehicle numbers. (The Keeper gave him those). It would probably take six hours to fill the paper work in. 'No, not worth the effort, waste of tax-payers' money.' The Keeper asked the policeman if he would book him if he wasn't wearing a seat belt. 'Oh yes.' But he's too busy for hare coursers!

\*\*\*

It's a sort of instinctive thing, keeping an eye on the turkeys. I'm always, subconsciously, checking where they are, because of their tendency to wander long distances. Then within that, I'm always counting them. So for a week now there's often been one missing. This is a clear indicator that they have started laying. Not clear at all, though, is where the nest is. They wander around the yard constantly, and amongst the hedgerows adjoining the yard. They are very secretive about where they lay, and we just don't have the time to mount a proper search. What we would like to do is collect the eggs as they are laid and eventually place some in an incubator and some under a turkey. It's serious stuff, we are talking Christmas dinners here. But eventually there are indicators of nest location. Why are those five magpies in that tree every morning? Why are those two carrion crows in the next tree? Simple, they are waiting for the turkey to get off her nest so they can have their breakfast. Next day we go to collect eggs, but there aren't any. The turkey hen is still on the nest, but she hasn't got a head. Mr Fox has been for breakfast as well.

\*\*\*

I've had some wheat straw to sell so I put it into the local paper small ads. On Saturday morning on my way to some rugby match in Cardiff, the phone goes. 'I see you've got some straw for sale.' *'Yes I have.'* 'How big are the bales?' *'I think they are eight feet in length.'* 'What sort of baler baled them up?' *'A Claas.'* 'How many strings are on each bale?' *'Six.'* 'Right, I know those balers, how many bales have you got to sell?' *'Fifty.'* 'Well I won't know until the end of the week how much I will want, is it good straw?' *'Very good.'* 'What was the weather like when you baled it?' *'It was one of the few sunny days.'* (As if I know that!) 'I don't usually have to buy straw but I had to bring the ewes in early because it was so wet and we've nearly used ours all up. I'll give you a ring in a week's time if I need some.' *'How much do you want?'* 'Oh, one bale should be plenty.' *'Right.'* All that took two bars off the battery of my mobile phone.

## APRIL 13TH 2013

Every year about five pairs of Canada geese make the one mile journey from a nearby lake to dispute the ownership of the tiny island in the pond in the field in front of our home. Apparently, if you are a Canada goose, this is a very desirable nesting site and the various pairs will do battle for several days until the victors make a nest and the rest, presumably, look elsewhere. This year was no exception: the geese arrived, the battles were fought, and the island claimed. Then the 'weather' turned up, and the geese vanished back to their lake. All this, year after year, has taken place in early February and it was the same this year. As I look out of the window, I can see more white than green, but today is the first of April and this morning the geese were back. There were nine of them, I don't know what the odd one is doing. It stands forlornly as it watches the others repeat their battles. All this takes place five weeks or so later than usual. And if someone asked you how late spring was this year, you'd probably say four or five weeks, which is exactly what the geese are saying, only probably more accurately. We can all learn a lot from nature if we take the trouble to look and listen.

Years ago we used to have peacocks that used to roost right in the top of a big ash tree. But if there was a lot of snow forecast, I would go to shut the hens up at night and the peacocks would be there on the perches amongst the hens. They might be there at night for a couple of days, possibly more than a week in a big snow. Then one night they wouldn't be there, I would shine my torch up into the ash tree and there they were right up at the top. Without fail, next morning a thaw would have set in. Just how clever is that?

*\*\*\**

We turned some heifers out about a month ago now. The grass fields I put them on had not carried any stock since the early autumn and there is plenty of the tussocky grass that I call 'picking'. They were the smallest heifers in a bunch and they've done really well since they've been out. They run a series of fields, one is at the top of the farm and exposed to the elements but they can run back down in to a small sheltered valley. But when there's ten inches of snow on the ground, the 'picking' is difficult to access so during the worst of the snow we've been taking them some nice spring barley straw to eat. That's what Stephen and I are doing now. We're putting the straw out in the valley where the heifers can eat it out of the wind. It's a beautiful sunny morning, the snow is pristine white in the sun. I'm driving the tractor and Stephen is throwing the straw off the back. Which is as it should be. It's been a real struggle on farms these last weeks but as I drive slowly up the valley I can't help being moved by the beauty of it all. A hundred yards ahead of me, four hares are chasing each other around in circles and flurries of snow kick up as they change direction. I have to remind myself how lucky I am to be living amongst all this. It all seems to make the hardships worthwhile. Stephen has thrown all the straw out now and the heifers are contentedly eating it, warmed by the sun. I stop so that he can get back on the tractor. 'Boy,' he says, 'Aren't we lucky to live in this countryside?' And there was me thinking I was the only one who noticed.

\*\*\*

These sheep farmers are something else. I'm on my Sunday afternoon feeding rounds and in the lane I meet a good friend of mine who keeps a lot of sheep. He, likewise, is going round his stock. We automatically pull up alongside each other and put the adjacent windows down so we can chat. We've not had a chance to chat for some time and it says a lot about the quiet area I live in that we can stop for 20 minutes and no other vehicle comes along. He keeps a lot of sheep: how many, no one knows except he, but I would guess it to be nearer 1,200 than 1,000. He's in the middle of his lambing and bears the unmistakeable seven-day stubble of someone whose days and nights are dominated by sheep and their care.

It is inevitable that our conversation centres around the weather, because the recent past weeks have tested us all. 'It's not so bad today,' he says, 'but I don't think I've ever been so cold as I was last week.' 'First time ever in my life I've had to put a hat and gloves on, it's cold enough today in the wind, but last week I struggled to keep going.'

He hasn't got hat or gloves on today but I can count his layers. I can see two thick shirts, waterproof top coat. That's three layers and each one of them is open down to his waist band! We start our respective engines and go to move on. As if at a signal, when his engine starts, the nasty little bitch he has with him in the cab, leaps onto his lap and tries to bite me through the window, but I've been there before and am already on the move. I give a bit of a shiver. Wouldn't want my belly exposed to the elements this weather.

***

It's a 6.30am phone call. I'm in the kitchen writing, which is a norm for me. It's a friend of mine. 'Could you do me a favour?' I say 'yes', but cautiously. 'I dropped the soap in the shower last night and hurt my back when I picked it up and now I can't put my socks on. Would you mind coming to put them on for me?' It's six miles away but what do you do? Having first ascertained that feet and socks will be clean, I set off and the job is soon done. I tell him that it would be better if he didn't take his socks off that night. 'But what if they want washing?' 'They'll get washed when you have a shower.' I washed my hands before I wrote this.

### April 20th 2013

A young dairy farmer phoned me the other day. He told me that his life's dream had been to milk his own cows, and that he had lived that dream for six years. He had now taken stock of his farming career thus far with his accountant and they had concluded that he had worked for six years for nothing and that they had only survived because they had lived on his wife's income as a teacher. The unpalatable truth is that they had paid the money that his wife had earned in order to milk cows seven days a week for six years. He is a really nice young man, not bitter about where he finds himself, but pragmatic enough to look at his situation in a realistic manner. He couldn't see anything that could happen on his farm that could improve his situation and so he had looked at possible alternatives.

One of these was to go and manage a herd of 500 cows in New Zealand. His two children are primary school age, his wife had been promised employment as a teacher and his income would be good. Because New Zealand cows calve on a seasonal basis they therefore stop milking all at the same time so he would have more time off to spend with his family. So how does he reconcile this with the dream of milking his own cows? Well he's done that for six years and paid for the privilege so milking someone else's cows and getting paid to do it is not that much of a step down, and the cows he is milking now probably belong to the bank anyway!

From the way he spoke I think he'll go. And it is quite a norm in New Zealand for herdsmen to build their own stake in a dairy herd so that they can move eventually to becoming dairy farmers in their own right. Just seems a shame that keen, motivated young families have to go to the other side of the world to fulfil their dreams.

***

Back in the snow, which was only a week ago here, I was putting silage in a trailer up at the top farm. I see the landlord's son walking by with his two dogs. He has two collies and goes everywhere with them. He's obviously been up to see his mother and is now walking home. At this time you could only get up to this top farm where she lives by tractor. When I drive the tractor home, there is a ball on the side of the track. I've seen his dogs playing with this ball often enough. It's the size of a proper football but has been punctured by canine teeth long ago. It sits there in the snow, it is bright red with black spots, a bit like a ladybird. I take my load home, tip it, and return for another.

The ball's still where it was before but there is a neat circle of six carrion crows stood studying it. I use the word studying quite deliberately, they probably saw the red of the ball first, red equals blood, equals food. So they have surrounded the 'food' but haven't worked out how to access it. The light is fading now and they will soon be at their roosts. I wonder if their empty stomachs made them dream of a potential feast just waiting for them in the morning. The word hate is not part of my life but I come close to hating carrion crows. I know they only do what carrion crows do, but it seems as if there's something inherently evil about them. If and when the spring turns up and I'm on tractor work in the fields, I will watch helpless as they methodically work the hedgerows, devouring eggs and fledglings. Must have a word with the Keeper.

\*\*\*

The turkeys are laying eggs and we've found their nest. It's down near the poultry sheds. The turkeys roost at night on an exposed fence on the yard and have roosted there throughout the bad weather. Given the multiple options available to them to roost in warmer, drier buildings, I have put them in the category of 'thick'. But down at the chicken sheds there are extractor fans at ground level that blow out some of the warm air inside the sheds. The stag turkey can be seen backed up to these fans, warming himself. Very much as I back up to the Rayburn in our kitchen to warm my bum. So he's not quite as thick as I thought!

\*\*\*

Egg laying is a protracted business and can take them all morning and it is only in the afternoons that the turkeys sally forth to take their tour of the yard. The stag is quite aggressive as he goes about, his feathers are always on display and one of my lorry drivers has complained successively to my son and to Stephen that the turkey has attacked him. Both of them, without hesitation, have told him that it is nothing to do with them. So he is looking out for me, and wants to have a 'word'. I'm quite looking forward to this word. It will be an opportunity for some of my best sarcasm. It's been a bad year on farms, the weather these last 12 months has tested us all. Then just when we thought we had been tested enough, March turned up. My sarcasm is all prepared. I will tell him about the implications of the weather on my business. I will tell him that 12 months ago my cows had been out at grass a month by now. I will tell him that at the end of February I had plenty of silage left and planned to sell some, but that now it has all been fed to hungry cows. The flourish of sarcasm that comes at the end of the speech will be to enquire of him where, in my list of trials and tribulations, does he think that his fear of a stag turkey comes?

## April 27th 2013

I always try to see the other person's point of view. If you can do that there is always a possibility of compromise. Providing, that is, that the other person is reasonable as well, which is not always the case. When I used to do a lot of farmer meetings, I would always sit down and try to pre-guess ten questions I might get asked and prepare a thoughtful answer. I hope it showed that I had given thought to their point of view. So this issue of culling lots of deer, as in a half million or a million, depending on what you read, leaves me completely baffled. It's a news story that has been about a few weeks. Nowhere have I seen anyone speaking out against this cull. There was a section on *Countryfile* devoted to it. We had the issues aired, the problems and the solution. Everyone seems to give a pragmatic shrug of the shoulders and accept that large numbers need to be shot. The justification seems to be that there are thousands too many, they don't have a predator, if we don't do something about it now, it will be twice the problem in just a few years' time.

Most farmers believe there are too many badgers about. Would someone please explain the difference? If I put my 'other person's hat' on and look at a badger, what do I see? Well I see a nocturnal animal with an attractive black and white face. I don't see many because when I am about in the daytime they are all well down a hole, out of sight. They are the nearest wild animal in this country to a panda and therefore just as adorable and if I want to see a panda I have to go to China or Edinburgh. If this person in the other hat is honest, they don't know much more about badgers.

With my farmer's hat on, I look at badgers and see too many. I see an aggressive predator that would, when I kept sheep, pounce on healthy lambs and unzip their rib cage and eat their organs. People with other hats on have never kept sheep (bit like my wife who has never, to my knowledge, mowed a lawn but knows more about it than me), they know more about sheep than farmers and will tell you that the lamb was ailing and would have died anyway. I for my part know full well that the lamb was fit and healthy but had made the mistake of going to sleep in the field next to the wood. As a farmer I know that badgers predate on the eggs and fledglings of ground-nesting birds such as the lapwing and skylark and will catch and eat very young leverets. I am told that my grandchildren's children are unlikely to ever see a hedgehog because badgers are very good at rolling them over and eating them and I also know that badgers play a part in the spread of TB. To illustrate just how real and scary the TB situation continues to be, I farm four blocks of land of varying sizes, so I have four sets of neighbours. At each location I have a neighbour whose stock has just gone down with TB.

So, still baffled, I turn to deer. They mostly live in woods but because of their agility, they can jump over most fences so that whatever the weather in the winter, they can always access food and therefore they flourish. I've yet to see the three in the wood here but it wouldn't make me lose too much sleep if they were eating my cows' grass. Though I do know farmers who struggled to get food to their sheep in the snow who found that they were feeding 50 deer as well. But largely they are a shy beautiful creature, not many animals more beautiful. The first film I can ever remember seeing was *Bambi*. Because I saw that film, I know that young deer are the most beautiful animals when young, that they have lovely curling eyelashes and they grow up to be either devoted mothers or noble stags who spend their time posing about on prominent rocky outcrops.

The deer that will be culled are going to be eaten, we are told, so that's OK then. I wouldn't be best pleased if I had invested in a deer farm and I was told that there were half a million more deer coming onto the market over the next few years. I know people who reckon they've eaten badger: less said about them the better. I just wish someone would explain it all to me.

\*\*\*

I've got this quite large shed that used to be a grain store. At some time someone lined the ceiling with polythene bags, possibly to prevent condensation. Feral pigeons and doves live in the space between roof and bags and you can hear them walking about. Sometimes a fledgling will fall through a hole and land on the muck amongst the cows I now keep there. They wobble about unsteadily and the cows give them a sniff and carefully walk around them. There's not much I can do to help them, I just hope their parents will come down to feed them. No need to worry in the end, because my old friends the carrion crows have eaten them all when I go around the cows in the evening. I know it's them because they are sitting on the gate waiting for some more.

## MAY 4TH 2013

Sometimes I will go to a meeting with a point in my mind that I want to make. Someone else turns up wanting to make the same point, makes it, and as long as it's made, that's fine by me. Some people are seemingly incapable of sorting out things like that and will make the same point again, different words perhaps, but the same point nonetheless. Sometimes there are three of them, but they will still repeat what has previously been said. I was at a meeting once where this sort of thing had gone on endlessly, the chairman wound the meeting up and his last words were, 'You didn't have a lot to say today, Roger.' 'Neither did a lot of other people but they still said it,' was my reply. Frustration can lead to sarcasm, well it does with me.

But I am going to break the rule here. Mainly because if someone or something amazes and intrigues you, well that fascination doesn't go away. You may remember the story of the abundance of cockerels we had here in the early autumn. Well, we disposed of them in the end. They were destined to make my son's fortune, but they didn't. There's a story that he tried to buy a pint in the pub in exchange for a cockerel but that's not strictly true – it was three cockerels. Anyway six of the cockerels were given to the Keeper who had a theory that they would act as a sort of focal point for his pheasants and stop them wandering, and four of these survived. The two that died were caught and mauled by spaniels, but I see the other four every day, now that we are about the fields on a daily basis. And here is what intrigues me, something that I still marvel at. They were reared in an intensive poultry shed along with 20,000 other chickens, cast out into a completely different environment to fend for themselves, and they did it successfully. I wonder at the basic instincts that enabled them to make that transition. They can't fly but the Keeper reckons they roost in the very tops of the tallest trees, well away from Mr Fox. He says they get up there by jumping from branch to branch, using the tree as if it were a ladder. How cold would it be at the top of a tree all night during the winter? And it wasn't just any old winter was it? How did they find food in a foot of snow? As I look at them every day, they are the picture of health, fine plumage, bright red combs. Good luck to them.

*** 

Over the last few days we have managed to drill our spring barley. Our first outing with the drill and power harrow was to drill five acres on our very top field. I'm not sure if we were in the mist or up in the clouds, but whatever, you couldn't see more than 50 yards so the result was the same. So how, I ask myself, after we've only been in the field about ten minutes, do ten buzzards turn up? How do they know what we are doing? I pondered this for some time: there they were, eating grubs and worms that we had disturbed and it intrigued me, just how had word got around. I couldn't see far, but could they see much further? Do they hear the tractors working? I still don't know and that makes it all the more interesting.

It was a fairly sunny day a few days later when we drilled 20 acres in another field, but we were under more pressure as the forecast wasn't good, and Stephen and I were booked to go on a bit of an outing. More of that later. Rooks were our most numerous companions, the usual buzzards and this time a heron. They keep very close to where you are working because that's where the worms and grubs are freshly disturbed. A tractor can be a mobile hide and bring you closer to wildlife than almost anything else, including birds that are only three or four yards away. It's a non-stop sort of a day with rain threatening and a thermos flask and 'bait' to keep you going. I always get a perverse sort of pleasure from throwing a banana skin out of a window. Every bird on the field will come up in turn to inspect it. The heron gave it a tentative peck or two. It must have thought it had something of the goldfish about it. Herons empty the garden ponds around here on a regular basis.

## MAY 18TH 2013

Stephen has been spreading muck on a field where we intend to grow stubble turnips for late summer grazing and some kale for next winter. Yesterday he was ploughing. This is hare country and during the day he spots four tiny leverets in the stubble he is ploughing. He stops each time he sees one and picks the leveret up and places it onto fresh ploughing, trying as he does so to place it amongst the clods of soil so that it has some cover. As he resumes his ploughing, he watches carefully and each time the carrion crows or the buzzards have found the leveret and carry it off before he gets back down the field.

***

The late spring has brought with it strange phenomena. Up the lane to the farm we rent, there are newly-planted quickthorn hedges either side. These hedges are now grown sufficiently to be layered next winter. New hedges like these usually show their first leaves before more mature hedges, but strangely, one hedge is about eight weeks ahead in leaf than the other although they are only 15 feet apart. Lots of people, well four or five, have remarked on this, but that's not enough for me, I have to work out why. And it's not that difficult. The lane runs north to south and there have been weeks of biting east winds. So it's simple: the one hedge has been sheltering the other. The easterly hedge has taken the brunt of the wind chill.

\*\*\*

The last planting has been going on this week, maize and potatoes. April was the driest month around here for the last 20 months as we had only about an inch of rain and crops look really stressed, especially winter wheat and rape. Large bare patches and even whole fields have had to be resown. All of this, and the new sowings, have been accompanied by clouds of dust. It actually rained last night, so that's a welcome end to that, but the conversations in the pub about planting potatoes and maize took an unexpected technical turn, when they started to discuss optimum soil temperatures for planting potatoes.

One farmer has a soil temperature thermometer and goes on and on endlessly about the soil temperatures in various of his fields. I don't take that much interest because there are other factors that are just as important, such as soil conditions and of course the date. It doesn't really matter if the soil temperature is right if there's been an inch of rain in the night, because you can forget about drilling anything for a couple of days. The consensus seems to come down to eight degrees being the optimum temperature but I can tell that interest in the subject is waning with everybody except the guy who has the thermometer.

But just when I think the subject has had its time, there is a new contribution. 'My old Grandad used to say that when the soil was ready you should ridge it all up ready for planting and then take your trousers down, your pants if you had any, then sit on the soil for five minutes on your bare bum. If after that time the cheeks of your arse are really cold, then it's also too cold for the potatoes.' There is a ready consensus that this method is far superior to anything a thermometer can tell you and someone goes to get another round, and the thermometer owner is clearly miffed.

**MAY 25TH 2013**

My wife has just had a new hip. She's had one before so I assume it's the other one. It's not been very handy for me, I've had to do some housework. Then there's all that visiting. The first thing I did was switch the Rayburn off which is always a contentious thing to do because she doesn't know how to get it going again. My next crisis is washing. I used to bring my washing down and drop it on the kitchen floor and it used to disappear and turn up a few days later, washed and ironed. This didn't happen this time for some reason, and one evening I managed to get the washing machine fired up and did it myself. I put it all out on the line and it's been out there in the rain now for four days. If we had a few hours of sun I would snatch it back in but I've got nowhere to dry it because I switched the Rayburn off. And I can't put that back on because that would be a sort of defeat. I didn't know life could be so complicated.

\*\*\*

Our charity fundraising at the pub is gathering momentum. I'm supposed to be chairman but there are three ladies on the committee with me so if I think I'm chairing meetings, I'm only kidding myself. We have about five minutes on what we will do next and then one of the ladies mentions someone's name and then we drift effortlessly into some gossip about that person or someone in their family. I don't try to stop this happening because it would be like trying to stop the tide coming in and King Canute tried that and it didn't work. I keep thinking that I will get something that I can write about but it's all much too salacious and libellous for me to repeat. Perhaps if I keep quiet and they forget I'm there I could get material for a column on women's problems so I'll try that and let you know what happens. Fairly soon we are having a 'savoury and sweet' competition. Last year we just had a sweet competition. I'm still miffed about it. I put a bottle of vodka in a green jelly but I didn't even get a honourable mention although mine was the only sweet that was completely eaten. I'm going to enter a savoury this year and I'm going to try to find something a bit different. Wonder where I could get some snails? Wonder what snails and vodka tastes like?

## JUNE 1ST 2013

I'm going up and down the field on the tractor, which is what you do when you're on the tractor; it's a bit like mowing a lawn only on a bigger scale, up and down until you are finished. A pair of carrion crows have appeared at the end of the field, busily eating something. I drive up to them when I get to the end. They are so busy with their prize that they don't move away and eventually my front wheel is only five yards away from them but they carry on eating. It's a tiny leveret that is providing breakfast. You know what I think about carrion crows and you also know what I think of leverets. But I mustn't be too hasty on this occasion. We cut the grass in the adjoining field yesterday, which is where the leveret will have come from: the crows didn't find it in this field or I would have seen it. They popped it over the hedge out of the grass field and are trying to eat it out of sight, without the competition of other crows and buzzards. They are probably not the villains that they would first appear. If the leveret was in the grass, it was probably killed by the mower so the crows are only 'clearing up'. This means that the leveret's death was probably down to me, not the crows. But I still don't like them. I'm sad that the leveret died, but the man who does the mowing says he saw lots of leverets and most of them ran away to safety.

## JUNE 9TH 2013

In the middle of May we experienced bitter cold and heavy, squally showers. I've got 18 young heifers out that I still feed some cake, to keep them topped up until 'spring' arrives. The irony of this is not lost on me, we are only a month from the longest day, when the nights will start drawing in and we haven't had spring yet! These heifers, when I get to them, are out grazing but they are grazing in a tight group, which is unusual. So I watch them a while; they haven't spotted me yet. The group is constantly moving in its tight pattern and I work out what they are doing. They all want to graze but they want to use their companions to take the cold wind off them, so they are constantly moving as they are alternately sheltered and then exposed to the wind. The best analogy I can give you is Emperor penguins shuffling constantly about on an Antarctic ice pack.

***

We're drilling turnips and kale this week, for food for the cows next winter. This is our biggest field, 40 acres, I'm on the power harrow going slowly back and forth, breaking up the fresh ploughing into a fine tilth. Stephen will start drilling the seed tomorrow, which is a speedier operation, so my work today will get me nicely ahead. My only companions thus far today are a few pairs of carrion crows, who search the fresh soil as I pass, for grubs and worms. Where are all the buzzards today? It's over an hour before one turns up, which is unusual. Within a few minutes there are 15 buzzards and there is a bit of jostling between the crows and the buzzards which includes some spectacular aerobatics, but they soon settle down as they realise that there's plenty of food for all. That's pretty much how it stays for a few hours. A lone seagull turns up but as far as I can tell, the other birds don't speak to it. As far as I can also tell, there are about five pairs of skylarks about. There's no way of knowing at what stage of their breeding cycle they are. If they had nests, they were probably ploughed in, and in that case I can only hope that they will make a fresh start. The irony is that just over the hedge is 20 acres of stubble deliberately left undisturbed for birds like skylarks. The other irony is that it must be a real struggle rearing a family of skylarks with all these buzzards and crows about.

Then there's a bit of drama. A pair of kites come over, effortlessly and with menace. Immediately the buzzards take to the air, not just the 15 that have been in this field, but buzzards in adjacent fields that I've not seen. This flock of buzzards soars higher and higher and I stop the tractor to try to count them. I count up to 30 and give up, I would guess there's over 50. Almost as soon as the buzzards are gone, my field is populated by rooks and jackdaws: it's as if they have kept away while the buzzards were there. But I'll be here today, probably for 12 hours, and although the views are spectacular and the bird-watching is much more interesting than the tractor radio, another diversion would be welcome.

And here it comes. A lone rambler. He comes into sight at the edge of the field and pauses there. It's our largest field and you can't see the next stile from either end. Because of the contours of the field, if you are standing right in the middle, you can't see the stiles at either end either. It's where I famously tried to spray out a footpath in some winter wheat. I started out at one end and sprayed to the middle, then I went to the other end and sprayed to the middle. The two lines were about 20 yards apart! I've never liked spraying out footpaths anyway but I had to join the paths together, so I sprayed out a rambler passing place and a lay by.

But what of my rambler? He consults his map and sets off across the field. I should add that if I am close enough I always stop the tractor to say 'Hello'. This completely throws some walkers who are expecting an angry farmer. It throws others who enjoy disputes and confrontation. But today's rambler is striding purposefully across the furrows. He leans forward, as he goes, against the wind, he has two ski poles to assist him and for a brief moment I can imagine Scott of the Antarctic leaning into a blizzard, dragging a sledge behind him.

He reaches the middle of the field and stops. He still can't see the next stile so he gets his map out. He studies this a while and then he gets out a compass. He uses both to plan his next move and then he packs both away and sets off, once again with purpose, at 90° to where the footpath actually goes. After about 400 yards he reaches the hedge, turns 90° again, repeats this when he reaches the next hedge and so after walking about 1,200 yards he gets back to the footpath again. The stile here is right next to the gateway. The gateway is 30 foot wide because the previous tenant used to drive his combine through it. The gates are wide open, it's how I got the tractor into the field. So instead of walking through this wide open space, he walks past it and climbs laboriously over the stile! Wonder what that was all about?

## June 15th 2013

I've been living here for 49 years. When I first came, there was a man living here in one of the cottages, who had been bailiff here for 43 years. Between us that was 92 years' experience. It's just one more man from the previous generation that I wish I'd spent more time with, but he moved away soon afterwards. He told me that the first job he did here when he started out was to create a tennis court. The garden here is on a gentle slope so they had to excavate soil and rock from the top of the slope and put it at the lower end in order to achieve a level area. Then they went into a ploughed field and worked it into a fine tilth and then shovelled up the top two or three inches and spread it out over the playing surface before they seeded it down. We've never used it for tennis, not serious tennis with a net, anyway. But we've had some fine football matches on it, and rugby and cricket. There is a reason why I bring this tennis court-sized lawn to your attention.

Moles. They've always been a problem, but never as bad as this year. I like mowing the lawn, I like the lawn to look nice. This is not compatible with moles. Our house is built on the rock, and half the lawn is on the rock with only three or four inches of soil to cover it. It's got to be the same soil that my man carried in by horse and cart all those years ago. So the moles work close to the surface and it's difficult to set a trap. A succession of friends have been here to try to catch a mole without success. It's so bad that I've had to rake out a quarter of it and reseed it. I did it one day in the heavy rain, assuming that the seed not buried would germinate anyway. When I finished I surveyed what I'd done (and the mess) and thought to myself, 'this is the worst it's ever been, it can't get worse than this'. But I was wrong. It could get worse. Our dogs roam free at night. They decided to lend a hand. They must have spent the night digging for moles. I've never seen anything like it.

### June 22nd 2013

On the farm track up near the top fields I pass an open gateway into a field of spring barley. The barley is only eight or nine inches high and just five yards from where I travel sits a hare. Over the two days she is never more than five yards from where I am travelling. I call her a 'she', because I am convinced that she has a leveret secreted away in the barley close at hand. As I go roaring past, with beacon flashing, the only sign she shows of having seen me is a flattening of the ears.

### June 29th 2013

A friend of mine works a day a week at a visitor centre near here, that used to be, many years ago, a big lead mining area. The other day she gets a visitor to the visitor centre. He tells her that he used to attend the village school where the centre is based. But that isn't all he tells her. He lived in a village about five miles away and his father was headmaster in the school there and his mother was a teacher in this one. His mother had taken him out of his father's school and made him go with her to her school, because, in her words, his father wasn't teaching him anything.

Seems that his father liked to take the boys in his charge on 'nature walks'. In fact, if it wasn't raining they went on a nature walk. And what walks they were. If the local foxhounds were meeting in the area they would follow them on foot, not just the autumn meets, but when they went cubbing early in the season. He particularly remembers going in the autumn to where there were some crabapple trees. His father was a proud Welshman and he would split the boys into two groups. Those with Welsh surnames on one side, English surnames on the other. The Welsh were the Welsh and the English were the Normans. This was as close as he could remember to getting a history lesson. His dad would be a member of the Welsh team and they would each pick ten crabapples and do battle, throwing the apples at each other as hard as they could. My friend, who is telling me this story, says 'I bet some people would find crabapple fights and taking school children hunting a bit 'iffy' these days.' Iffy! Iffy wouldn't come anywhere near it!

*\*\*\**

I had occasion to be in a farmhouse the other day where I'd not been before. There was an old Granddad there in his late nineties. What a man for his age. Dressed up smart, collar and tie, smart jacket and trousers, the whole outfit colour coordinated, conversation excellent. As we sat there at the kitchen table drinking a cup of tea, I thought to myself, if I could be as fit and as smart as that at that age, well I wouldn't mind getting that old. Then he took his false teeth out, put them on the table, and started cleaning them with his pen knife. Perhaps not.

### JULY 6TH 2013

Red kites are scavengers. We all know this because they tell us on nature programmes. Because they are scavengers their proliferation and spread is one of the wildlife success stories of our day. So that's all OK, because 'they' say so. But meanwhile, back on my farm, I'm thinking about hares. On the land I rent a mile away, although there is a lot of cover everywhere and as crops grow, there are hares everywhere. Here at home, on 80 acres, we've not seen a hare for years, but now there's been two hares about for a couple of years, and I've been really pleased about this. So yesterday Stephen was mowing about five acres of grass. It's out of the cows' grazing area but the grass has got ahead of the cows and it's best to take our area out with the mower so that the regrowth is the sort of grass we need to produce milk.

He sees two tiny leverets in the grass that he has disturbed so he stops the tractor, catches one leveret, and carries it 50 yards to a different patch of grass and safety. As he is returning to catch the other leveret, two kites turn up as if from nowhere. Before he can get back to the second leveret, it is snatched up by a kite and is being carried away up to the wood, although it has been picked up only about five yards from the tractor. He turns to see what has happened to the other kite just in time to see it carrying 'his' leveret away as well. Of course all this is my fault for getting the mower out. And kites continue to spread their range further and further, and do more and more damage. But it's OK, they tell us, because they only scavenge for food. But do the kites know that they're supposed to be scavengers?

*** 

It's a beautiful Sunday evening and my son is having a barbecue with his family in the garden. A hot air balloon goes by but no one takes much notice of it. Ten minutes later he can hear our cows bawling so he goes to investigate. The cows should be grazing half a mile away but they are all in the buildings and in a frenzy. It's chaos. Gates broken, fences down, cows everywhere. Order is eventually restored but when he milks next morning he finds one cow had lost the end of one of her teats. So you have the challenge of a quarter of an udder full of milk and the only way to get the milk out is through an open painful wound. The two are incompatible.

It's a job for the vet. Infection of the affected quarter is almost inevitable so the cow goes on antibiotics and the milk out of her other quarters is unsaleable. And so it goes on, and the cow will probably have to go. If you added it all up and bought another cow, you could quite easily be £1,000 out of pocket. We've tried to identify the hot air balloon but that's not proving easy and no one's putting their hands up. It might be a rare case, but nevertheless it's one more case of society 'dumping' on farmers. Two years ago we lost a cow in strange circumstance. We don't always do post mortems on dead animals, I've yet to see a post mortem bring an animal back to life. But this was a strange death and made me curious. Turns out the cow was killed by eating wire on one of those magic lanterns. Why should I have to pay for that sort of irresponsibility?

## JULY 13TH 2013

It's just about still dark when I fetch the cows for morning milking; most have already drifted home but there's always a few stragglers to get in. I set off down the track, there was no sign of Mert or the corgi this morning but I'm not alone. There are four foxes on the track in front of me. They are going at the same speed as me, which isn't very fast, and they are maintaining a distance between us of about 20 yards. So what's that all about? I know what it's about. These are town foxes, newly dumped. If they were country foxes I wouldn't even have seen them, they would have heard me coming and have gone out of sight long ago. Our little procession gets to the field gate where there are nine cows still in the field. I stop there and call the cows. They come quite quickly, they don't realise that Mert isn't with me. Because I've stopped, the foxes stop. One of the cheeky beggars sits down and yawns. But no one catches foxes in towns and releases them in the countryside do they? These must have just been extra-friendly.

*** 

I shut the turkeys up every night with the help of a big stick, and when I let them out next morning I just unlatch the gate and run. I was doing just that yesterday, went quickly around the corner and strolled off around a couple of parked cars for safety. We used to have a trick when we were children of hitting someone on the back of the knee when they weren't looking and the knee would buckle and they might go down. Turkeys know this trick as well. I was hit at the back of both knees by this huge turkey and went down in a heap. But not for long, because he was on to me all beak and talons. I managed to escape, but not with much dignity.

***

I've spent two days at the NEC at what they call the Livestock Event but what used to be called the Dairy Event. Most of my time is spent on a stand put there by my dairy cooperative, talking to members. We've got a new soft cheese product out called 'Quark' which our members can sample. It's a bit like a yoghurt to look at but can be used to cook with or eaten with a flavouring on its own or on sweets or cereal. It's so healthy that it's the only dairy product that gets five green traffic lights with the government's new food classification. Anyway, enough of these commercial breaks, I only mentioned it because I think it's good to see British dairy farmers, who own their own company, coming out with innovative products. But there's a fair gang of us on the stand and we come from all over the UK and we get together in the evening to eat and socialise.

There's a farmer from Scotland telling us a story and he's getting a fair bit of stick about his broad Scots accent. He's already told a couple of us the story and we know it's got a happy ending so it's OK to tease him a bit. He tells us that at home he's got a border terrier called Milly and Milly likes to go up in the woods rabbiting. But it doesn't come out like that. It comes out like 'Mae wee dog Milly is away up the glen chasing yon beasties.' Anyway we all fall about laughing and I interpret some of it for the others, but not necessarily accurately! Anyway, one day Milly doesn't return, the family are distraught that Milly is missing and enlist the help of someone who has a reputation for being able to find dogs stuck down holes. But he spends hours with his head down holes with no success. When he goes he tells the family that the only comfort he can give them is that he has heard of dogs reappearing six weeks later, alive, but not necessarily very well. They have lost so much weight that they have finally been thin enough to wriggle free from the roots where they are stuck. It's not a lot of comfort, that story, it's got to be a long shot. But five weeks and six days after she disappeared, an emaciated Milly struggles up the yard and back to her family. Happy endings don't get much better than that.

## JULY 20TH 2013

People – the public at large – don't like farmers spraying. Spraying is chemicals, chemicals is bad, chemicals is skull-and-crossbones sort of stuff. I can remember a time when spraying crops as a routine was not that widespread. I can remember, now that I am warming to this theme, coming second in a Young Farmers county rally in Monmouthshire for calibrating a sprayer and the tractor (like most tractors at the time) didn't have a rev counter on it and you had to step out a distance and put the throttle lever at a pencil mark on the bonnet and see how long it took to travel that distance and work out your mph! Who needs GPS and satellites? I can remember when most cereal crops were spring-sown and consequently harvested later in the season. Later means shorter days and heavy dews and damp crops. And damp crops, full of weeds that haven't been sprayed, are difficult to combine, and I can remember drivers on old Massey combines struggling to keep going because the weeds would block the combine.

And the evening would be getting chilly and they didn't have cabs and they would be wearing big overcoats to keep warm and they were so tired when they got home they didn't bother to shave so the stubble on their faces was black with the filth that had come up off the weeds all day. Then we had sprays and we didn't have weeds, so all that hardship ended.

That's a lot how farming evolved. There was never a time when two old boys leaned over the wall of a traditional pig sty scratching a sow's back and one of them, with a flash of inspiration, a eureka moment, said, 'I know what we'll do. We'll make a pen out of tubular steel and put it on concrete and put that sow in it so that she can just, but only just, stand up and lie down. We'll feed her at one end and clear the muck from the other and we'll have her where we want her. And we'll call it a sow crate.' He didn't think that, because sow crates were a long way down the road and sow crates were a particular pendulum that swung too far, especially for people in this country, though they don't seem so worried about buying cheap bacon from abroad which started its life like that! No, our two old boys didn't think that, but subconsciously they probably wished it. Why? Well, our sow probably stood on a cobbled floor and she probably spent a lot of her time, with her powerful snout, trying to loosen just one cobble stone. If she succeeded with that all the other stones would be ripped up in quick succession. When she sought a diversion she would try to loosen the bricks of her sty. The door of the sty would be an obvious target as a route for escape.

When she had piglets they would be born in the sty. As she lay there farrowing each emerging piglet would struggle, umbilical cord in tow, towards her grunts. When it eventually got to her head there was always a chance she would eat it. Those piglets that didn't get eaten, the sow would do her best to lie on, when she flopped down to suckle. Any human who tried to intervene and save piglets from being eaten or crushed had a fair chance of being eaten as well. I used to keep 100 sows many years ago and I would have given them up rather than confine them in a sow crate. But I have to say, with honesty, that there would be individual sows that I would cheerfully have put in a crate because of their behaviour. It's all about how systems evolve and sometimes they go one step too far and have to be brought back a bit.

Anyway, I was supposed to be talking about spraying. How you got me from spraying to pigs, I've no idea. We don't have a sprayer, our spraying is done by a contractor. We spray our cereals to kill weed, for the obvious reasons previously mentioned. And that's usually it. We rarely use pesticides. The only time we do is when we grow turnips or kale. The plants can come up to the two leaf stage and be eaten by something called a flea beetle.

This is a critical time. If you have a dry spell, the beetle will decimate your crop, but get a timely shower and the crop will grow away faster than any beetle can eat it. But we do 'spot' spray thistles and nettles. I used to do this with a knapsack sprayer on my back but I'm getting too old to carry one of those around all day so we've moved on to an electric sprayer that sits in the back of the truck and all I do is drive around in four wheel drive with a hand lance out of the window and zap thistles and nettles as I drive slowly past. It's easy, it's comfortable, the radio is on, there's nothing to it. So your driver's window is open and you sometimes get the wrong side of the wind and you get sprayed as well, so what? And you get some in your mouth and some in your eyes, but it's better than having five gallons on your back slopping about. Or is it? No, it's not. You burn your throat and you burn your lungs and you spend two days in the house feeling sorry for yourself. In fact you feel so poorly you don't even go to the pub on Saturday night. And that's *really* poorly.

## July 27th 2013

Most of the fields I rent have a six metre conservation margin around the outside. It's designed to be a wildlife corridor. In July we are required to cut two metres of this margin, the two metres that are adjacent to the crop. I've been cutting the two metres this week which is a bit like driving into a jungle because by July, the weeds and grasses are as high as the tractor bonnet. It's just a bit strange because although the fields themselves may be in grass for silage, cereals, or fodder roots, they are largely undisturbed until I go driving around the outside.

At this time of year there is plenty of cover for wildlife, a crop in every field, but I see probably 20 hares moving away as I approached. They must have had a good breeding season because any hares that have any sense will be in the woods in the shade. I see lots and lots of pheasants. There are hundreds of pheasants left on the ground after the shooting season, most of them hens, and as the months have progressed I have watched them go through the cycle of breeding, laying, sitting eggs and then what? Nothing. I have yet to see a hen pheasant with live chicks! I suspect that buzzards, kites, ravens, magpies, carrion crows, badgers, foxes, you name it, have had all their eggs, and if one chick were to make it alive, it would soon go the same way. So it's not difficult to imagine the effect they have on wild birds.

My journey has taken me to the fields where the dry cows

are grazing. I go into their field and see there are a few nettles and thistles that need my attention but not as much attention as the cows. There are 36 cows in this group and it's the time of year when they can get what is called summer mastitis. It's caused by flies infecting the udder via the teat end and it can flare up very quickly, is difficult to treat and can have fatal consequences. As I approach, all the cows are lying down in a fairly tight group and the experienced eye can soon tell if something is wrong. I can see something white lying at the back end of a cow. It looks as if the cow has had a white calf but the calf doesn't look right, it could be dead. But it's not a calf, it's one of those white cockerels we let loose in the woods last year. He's stretched out in the sun taking his ease and he's actually leaning on the cow for extra comfort.

\*\*\*

We've all heard of the Hay Festival. Never been myself but assume it's a sort of literary bonanza held in a lovely town that boasts a lot of bookshops. Been to Hay lots of times, it's a nice place to travel to as well as a nice place to arrive at. My favourite journey is to go to the north of Abergavenny and up the Llanthony valley. But there's been another 'Hay' festival these last couple of weeks. I've been on holiday to Scotland this last week, travelled miles and miles, and I've never ever seen so much hay made. It must be thousands and thousands of acres. Hay that has been made on the back of remarkable hot weather and with the confidence of a good weather forecast. Most of the hay made was originally intended to be silage and most of that would have been plastic wrapped round bales, but the timely sunshine has saved the cost of the plastic wrap and once you've bought it you have to pay to get rid of it, so it's a double win-win.

Personally I'm a hay-making sceptic. Why would you risk such a high percentage of a field's annual output in the UK climate? But even I have made hay. Like the rest, I cut a couple of fields to make silage bales but I intended to only leave it to wilt a couple of days to make a sort of half-silage half-hay product that we call haylage. But we don't have any hay-making kit, for the reasons I have stated, so by the time we had cut the grass and waited to borrow something to turn it, it had by good luck turned into hay anyway. And very good hay at that. This has blown my no hay-making theory out of the water. Previously I had considered hay-making to be a bit like going to Blackpool:you should only do it once every ten years, if only to remind yourself that you shouldn't do it more often. Farmers' morale was in need of a spell of harvest weather like we've just had, but farmers are farmers, and it won't be long before they are all crying out for some rain.

***

When I was a little boy, which is a long time ago, I used to fit a piece of cardboard to the spokes of my bike with a clothes peg so that the cardboard rubbed the spokes as the wheel went around. This would make a noise as you rode your bike which would be as near as any of us would get to the noise of an engine for many years to come. During the years of waiting for an engine, I discovered girls, so that the first engine, when it arrived, wasn't quite as important as I thought it would be.

There's always something that puts me in mind of these reminiscences and today it's car rallies. We've had our second one around here at the weekend. They are always on a Saturday night, or rather early Sunday morning.

The organisers spend a lot of time telling people what's going to happen. For our part, we have 30 acres of grazing land with an unfenced council road through it, and we always move the cows away for the night so that the rally don't have to open and shut gates. Each to his own, live and let live, it's not a problem to me. But why do they have to make so much noise? If it's all to do with driving skills, why can't they have the same exhausts as the rest of us? Why must we have to hear every gear change they make within a mile of where I am trying to sleep? That's not to mention the throaty roar of the double dip on the throttle just before they start off.

I actually know what it's about. It's about boys and their noisy toys. They are no different to me fixing cardboard onto my spokes all those years ago, and where was the harm in that?

## AUGUST 3RD 2013

I've been thinking a lot about hen pheasants and their conspicuous lack of success at rearing a brood of chicks out there in the wild. Now, because I can always see both sides of everything, I will be the first one to put my hand up and say that as far as I can tell, pheasants are not very good mothers, and that some of the lack of success is almost surely down to them. And also, in the grand scheme of things, it doesn't really matter if they are successful or not because their breeding activities are of no significance – because 'new' pheasants are readily, commercially, available. The Keeper tells me that he has 8,000 new poults due to arrive shortly. But their lack of breeding success is an important benchmark, in my opinion, because of the activities of predators. It's a very visible indicator.

You see the pheasants in the spring, you see the cocks, each with its own territory. You see the hen pheasants as they stumble off their nests once a day, their legs stiff from all that sitting on eggs, looking for a quick feed before they go back to their eggs. And now you see the hen pheasants out and about and you don't see any pheasant chicks. It's the same with partridge, fewer of them, but more wily, but still no chicks. If I am right, and I'm sure I am because lots of farmers are telling me the same story, it seems obvious to me that predation is having this serious impact on the breeding activities of game birds, an impact that I concede doesn't really matter. But it therefore seems reasonable to assume that song birds, particularly ground-nesting birds, are suffering in a similar fashion, and that really does matter.

The RSPB are currently running an excellent advertisement about the importance of caring for nature. It's a very clear message: if you don't care for it, you won't have it. I endorse that to the full. What I struggle to understand is that, if it is commendable to help nature at one end of the scale, why is it so difficult to see that out there, in the fields and woods, there is another section of nature that needs some help too? And if the RSPB advertisement is highlighting the need to help nature now to have a nature in the future, what sort of legacy are we leaving for future generations if 'we', (not me who can see what is going on, but 'they', the people who should be doing something about it), ignore the ravages wreaked by predators today. I was at a funeral yesterday and at the wake afterwards a group of farmers were discussing the very same subject. No prompting from me, I just happened to join the conversation after it had started. And the conversation was very much as I have outlined thus far. Ravens, crows, magpies, kites, buzzards, foxes, badgers: they were all named. One farmer was saying how he had cleared a field for silage and the next day he had counted 59 buzzards and kites looking for food. 'What's the sense in that?' he asked, 'Ten of each would be plenty, then some of these little birds would have a chance.' Such common sense.

***

Tractors are funny things. We recently swapped a 26-year-old tractor for a 13-year-old. This turned out to be an inspirational decision. The old tractor was really rough, and who wants to spend long hours in very hot weather, on a rough tractor? Then there's a plus. The 'new' tractor has air conditioning! But the weather got hotter and hotter so that there came a time, one midday, when I thought to myself, 'This air conditioning isn't working very well.' Then I got off the tractor to open a gate and stepped into furnace conditions and I realised that the air conditioning was working very well indeed! But more modern tractors are not, if you will pardon the pun, all sunshine.

Next day I go back into a hay field with the hay-turner on the back of the tractor and the hydraulics on the back will not go down. Older tractors had levers, levers go up and down, implements usually do what levers tell them. This tractor just has a switch. Switch it where you will, nothing happens. We are now in the territory of sensors and solenoids. I sit there for an hour on the phone speaking to mechanics of varying superiority who, with barely concealed patronisation, guide me through a sequence of instructions. 'Has the dial said this, has the dial said that? Right now it's on the ground.' 'No it's not.' 'We'd better come out to it with the lap top.' 'When?' 'We are really busy, could be two days, oh it's Sunday tomorrow, three days' time.' As it happened the weather was so good it didn't matter, but it could have done. That's all a week ago now and we still can't use the tractor. They've been out to it a couple of times. 'We think it's the main control box.' So they take the control box off and plug it into a similar tractor. Works perfectly. 'Could be a wire.' Right.

## August 10th 2013

Gamekeepers are always about. If you went sneaking about their woods, and then stopped suddenly, then one of them would surely bump into the back of you. But they've been through their quiet months and now they are very visible. July is the month most of them get their new pheasant poults and before they arrive, they have to repair their pheasant pens. These pens are always in a wood somewhere, strategically placed so that when the shooting season arrives, the pheasant travel every day to a game crop or to somewhere where they are fed a distance away, so that on shooting days they will instinctively fly back to these rearing pens and 'home', and over the 'guns' that are doing the shooting. The rearing pens will have five foot wire netting around them and some of the poults will eventually fly out but that's OK, based on the theory that if they are big enough, they are old enough. Perhaps the most important part of the rearing pens is the electrified wire that runs right around the outside, just a few inches off the ground. This is to keep out foxes but more importantly badgers.

Mr Badger takes a lot of keeping out of these pens and if he gets in, he will kill a few poults, but the danger is that he will get in and leave a hole in the fence for the fox who will get in through the same hole and kill all the poults, even if there are a few hundred. And that is the other element of the Keepers' July activity. Mending the pens by day and out at nights with their rifles after foxes. Any litters of cubs that have been reared will be out and about at night now, looking to establish territories – and powerful lamps and powerful rifles will be their end. Some shoots will use a few hounds to flush out foxes from the densest woods so that they can shoot them as they break cover.

***

We've got a fairly steep field that is in a stewardship scheme where we can't spray or fertilise. It's got a few too many nettles and thistles on it for my liking. These didn't get cut back or topped last year because whenever it was dry enough to top, there was plenty of other work that needed doing. But I did get on it this year in the dry weather and was dismayed at the size of some of the clumps of nettles. One clump was huge and high so I dropped down a couple of gears as I cut it off. You never know what these nettles will conceal: could be an old tree stump or just a branch that has broken off. Either could damage my topper so it's best to proceed with caution. What I didn't expect was an explosion. Half way through and there's a white eruption. It's two of those cockerels we let loose last year. They must have gone into the nettles to seek shade. From the way they took off I must have frightened the life out of them. But not as much life as they frightened out of me!

***

The Keeper hasn't yet accounted for all the foxes. Over the next days, two separate piles of white feathers tell a sorry tale. I can't continually provide you with stories with happy endings and the demise of these two cockerels leaves me a bit sad. It's not so much that I miss them, it's just that I admired them so much for having gone from an intensive rearing environment to surviving out in the wild successfully, survival that included one of the worst winters we've had for years. If they'd only managed another week they would have probably got into a rearing pen with the new pheasant poults and enjoyed the same protection that they get! So there's only one cockerel left now and he goes about with the dry cows for company. I just hope he gets well up a tree at night.

*** 

I was at an agricultural show yesterday. A tremendous show in tremendous weather. 'They', the people who organise it, reckon it's the biggest and best one-day show in the UK. And to be fair, it's got to be up there with the best. But I used to go to a lot of shows and know of several that have awarded themselves that accolade. I can think of half a dozen fish and chip shops that reckon they are the 'best in Wales' and say so on their signs. My favourite is one in the valleys called 'A fish called Rhondda'. But at this show a friend of mine was much taken with a beautifully-made traditional shepherd hut on wheels. So taken with it that she thought her husband should buy her one. It was very quickly turned into a discussion of pros and cons, she elaborating on the merits and usefulness of owning one, merits that included the option of solar panels that would power laptops and television, which bring new meaning to the word traditional. He for his part made a big thing of the price tag of £12,000. Which I thought a bit churlish. If my wife wanted to go and live in a traditional shepherd's hut I'd gladly buy her one. Providing I could choose where it was sited.

### AUGUST 17TH 2013

For many months now I have tried to bring to you something of the despair, anguish, trauma, the list of appropriate nouns is endless, that comes with an animal being identified as carrying TB. And I've tried to explain that those difficulties come way beyond anything that statistics will tell you. And all that I have tried to tell you has been second-hand. The experiences of others. Well, that will not be the case any more. From now on it will all be my own experiences, because last week we had a cow that reacted to the TB test. First one for 20 or 30 years. We've still not come to terms with its implications. The most important of these, that we can't sell any stock, unless for slaughter, until we have had two clear herd tests. These tests will occur at 60-day intervals, so the best, the very best-case scenario, for us, is to be able to sell stock in four months' time. An important part of our income is the sale of calves. Beef cross calves and dairy bull calves, we have over 30 heifers and 50 cows to calve in the next two months. The calves for sale normally go at a month old. We either slaughter these or keep them to sell at a later date. My very natural inclination is to keep them. But we don't have the resources for that, the housing or the food, so we will have to shuffle around to achieve the former, and buy the latter. We lose calf income and incur extra expense. We will need the help of the bank manager.

The most frustrating aspect of all this is that the impact I have described will negate a whole year's work and the best we can hope for is a break-even situation at the year's end, which means, as the bank manager will quickly tell you, that you have in fact worked for 12 months for nothing. Should there be any further negatives, such as adverse weather, we will make a loss. And there is nothing I can do to stop it happening again. My stock do not come into contact with anyone else's stock. I bought some heifers last spring but they've not been within two miles of this farm and they were all clear anyway. Is it any wonder farmers get angry and frustrated at the lack of progress on this issue? You've never seen me angry and frustrated have you? I suspect you will.

## August 24th 2013

I've been talking to a farmer I know, he doesn't live around here, which will be some relief to those that do. He milks about 1,200 cows on four units. I told him about my TB issue, so he told me about his. His farms have been closed down now for two years. He puts every cow to a dairy bull and as a result, every male calf born in the last two years has had to go to slaughter, which has got to be over 1,000 baby calves! Another number that will not be included in the statistics! By the same token he has had to keep every heifer calf born and rear it, which is OK, but only up to a point. Because the heifers are starting to calve now. Obviously some were needed as replacements in his own herd, but not all that number. No one wants to buy them, why would they, they have TB? He can't sell them anyway, because he is closed down. So he is having to take them into his own herd. But he only has room and feed for 1,200 so to make it all work and fit in, he is having to send healthy productive older cows for slaughter to make room for the calving heifers. It's a choice he has to make, a choice forced on him, sending cows he has bred himself to the abattoir, two or three years before their time. Puts my problems into perspective.

\*\*\*

Life can take you into strange places but never before has it taken me to a fashion show. We try to have an event a month for the charity we 'do' at the pub and last month one of our little committee organised a sort of wedding fair in a local hall. I didn't get involved in the organisation myself and if I'm honest I couldn't quite see how they were going to make money, but it was incumbent on me as chairman to go, so I did. There were stalls down the sides of the room selling everything from hat hire, to cakes, to balloons. There was a catwalk down the middle where local models paraded in wedding dresses and mother of the bride outfits. No anorexic super models round here! Some of the brides were belters *(note to self: sexist comment?)* They even had a gown for the pregnant bride, which I couldn't understand, thought you had to be married to get pregnant. My abiding memory is of the faces of the women and girls sitting opposite me as they gazed at each model, looks of absolute rapture. Turns out we made nearly £1,000 when the money is all in, which just shows how much I know about it.

### August 31st 2013

We're just starting our calving season. All our dry cows go away from home and at this time of year graze a string of four fields that start as a narrow valley and wind their way up to the very top of our 'away' land. There's seven to come home that are close to calving. One is so close to calving that she calved yesterday, so there's seven cows and a calf to fetch. My son goes to sort them out, they are in a bunch of 40. He always goes on his own, he's not away long and he is always successful. We others don't know how he does it, we think he must be one of those cow whisperers.

So Stephen and I are sent to do the haulage and we decide we've got two loads of three cows and the cow and its calf on their own. What we do is pull in off the road with the trailer and back the trailer tight up to the roadside gate so that there is just room for the cattle to turn into the trailer but we've got a corner to drive them into.

Previous experiences have taught us that there are two weak points. We need a spare light gate to help us if a cow should be fractious and there's a gate tied to the fence by the road in case they try to jump out instead of going into the trailer. We put the trailer into the right place and find that both of these gates have gone. Too late now, the cows are sorted, so we push the first three towards the trailer and in they go. Same with the next three, we're well pleased, the cows are fit and well and full of themselves.

After lunch it's just the cow and calf, easy, the calf strolls into the trailer and its mother puts her head over the fence and climbs over it. We put the calf onto the road to try to draw the cow back but the calf legs it up the road after its mother. I send Stephen in pursuit as my cow chasing days are long gone. She and the calf go half a mile up the road and into a 40 acre corn field. Two hours later they stroll nonchalantly into the trailer and Stephen thinks he is about to die. That's two hours extra work just because someone has nicked two gates there, so we'll have to do some extra fencing. You never know who is about but I suspect they drive Transit trucks with chrome wheels.

### September 7th 2013

We were asked out for supper by my niece who wanted a favour off me. Which sounds a bit harsh because she's a bit of a favourite of mine so the favour was an excuse really for hospitality gladly given. There were another couple there, neighbours, a really nice couple who happened to be vegetarians. That in itself was not an issue because they weren't 'in your face' about it. In fact it only came up in conversation when we were talking about farming and the man said that if he were ever to keep animals commercially he would choose deer, because they would be slaughtered on farm and would not therefore have to endure the terror of a ride to an abattoir, all said quite gently, as a genuine point of view. I for my part, equally gently, pointed out that you mustn't attribute human perceptions to animals and that just because you know and the lorry driver knows that the lorry is going to an abattoir, the animals don't know at all. I do concede that animals have to travel too far to be slaughtered because many local facilities have closed, largely because of the financial burden of bureaucracy on them. So that's that.

But an hour later the conversation has moved on, as conversations do, to pets. Turns out he has all sorts of pets, all rescued, including two pythons. Turns out he feeds the pythons live rats. He tells us he drops the live rats into the glass tank, OK, I know it is called a vivarium, just don't want to appear pretentious. He drops the rats and runs out so he doesn't see them eaten. Sheep in a lorry, rats trying to escape from a python? Don't think he'd thought all this through.

***

We have a lot of trees around our house. They are a source of great pleasure and highly valued. That is, valued as fine upstanding trees and not as in any value they might have on the back of a lorry. So it's a source of sadness if one dies. And one has. A really fine copper beech. It's something I wonder at, how did such a huge organism just stop growing? It's right next to the lane up to the farm and I decide it's better for me to get it down than for the time to be of its own choosing, when it might come down across the lane, block it, and do damage. I asked a timber man to have a look last year. We had one of those sharp intakes of breath, 'Ooh that's tricky,' and he points out the adjacent power lines and telephone wires. 'We'll have to get a tree surgeon in and he'll have to climb the tree and take it down bough by bough, you're talking at least £500.' Now that's a lot of money. But then it isn't. I wouldn't go up that tree, dangling on a rope with a chainsaw in my hand, for ten times the amount, (and possibly ten times that). I don't do heights, me. Twelve months later and I've still got my £500 and the tree is still standing. I phone a friend of mine who has worked in the woods all his life to come and have a go at catching my moles on the lawn. I'm getting desperate now, and he wants to know what I will do about the dead tree. So I tell him the story and he says 'I've got a friend who will fetch it down for you without all that climbing nonsense, we'll do it next week.' And fair play, on the due day, they turn up. I'm not that inspired to be honest, I'm the oldest of the three of us, but not by a lot, and we're all on our pensions. But I don't really count because I'm keeping well away, felling trees can be scary stuff.

They want Stephen to put them up in the bucket of the telescopic loader so they can trim off the ends of the boughs that are near the wires, one has his ordinary chainsaw and the other a chainsaw fitted with an extension bar. To be honest as the two of them go up in the bucket it's a bit like the *Last of The Summer Wine* going up to heaven. They do their bit of pruning in ten minutes and another ten minutes later they tell Stephen 'She's off now' and he gives it a tug with the tractor on a rope they've attached half way up, and 'she' is down. Lying neatly alongside one fence and within inches of another and not a wire touched.

So we come to the money side of it all. That's no big deal either. They want £200 which as you may guess is a lot less than £500. But then they want to buy the tree to sell as firewood and offer me £180. I tell them I was hoping for £200 for it, so that's the deal. A couple of people have been quite indignant, 'Why did you cut that lovely copper beech down?' 'Because it was dead.' 'Are you sure it was dead?' 'Well it hasn't had any leaves on it for two years, which is always a good sign.' I'm a bit indignant as well, I thought they knew me better than that. I would not cut a live tree down. So why did it die? I asked the two who felled it and they pointed to the bark peeling off and thought it might have been touched by lightning, but when it's down we can see that the trunk is almost all rotten and there probably wasn't enough live wood left to keep it going. It had had its time. A new copper beech will go in in the autumn.

**SEPTEMBER 14 2013**

This is one of our calving seasons: there are cows calving every day now. Thus far, we have eight dairy heifers that we will keep and nine beef calves that we would normally sell at a month old but cannot because of our TB shut-down. There's milk powder to buy for the beef calves. They live in the shed where we usually rear the heifers. Slowly but surely, TB impacts on everything we do and how we do it, and it will only get worse.

\*\*\*

Yesterday we were putting in some grass seeds in a former field of wheat. I was working the field down with the power harrow and Stephen was an hour behind me with the drill. Two buzzards and four carrion crows turn up as soon as I start moving the soil about but they are not there long because here come five red kites. Kites seem to be at the top of the pecking order in many ways. I think what happens is that there are grubs and the like having a bit of a snooze just below the surface of the soil and suddenly this wildly rotating machine comes by, shakes everything up in its path, and your grubs find themselves lying on the surface with probably the worst headache in their lives.

Before they can gather their bodily resources together and burrow down back to safety, birds of prey delicately harvest them. In addition, what a delicate harvest it is. There's a strong wind blowing and these kites hang almost motionless on that wind, they seem to toy with it, they allow the tractor to get as close as five or ten yards so you can see all their beautiful plumage, they are well aware that the closer they get, the more grubs they get.

***

At this time of year, the pheasant poults have largely escaped from the confines of the rearing pen and are starting to feel their feet and explore the world. They travel about in quite large groups and seem to have a fascination with roads. It is not uncommon to go around a corner of a lane and find a group of 20 or 30, often several times that number, just stood in the road. They might be big on exploration but they are not big on the Highway Code, and their reaction to tractor or truck is usually stupidity. It drives the Keeper demented because he works full time on a farm and is busy harvesting at this time of year and has to do his Keepering duties very, very early in the morning and sometimes late at night, if he's finished work in time. The agony for him is if the pheasants run in front of a vehicle for a long way, which they often do, a run which may take them off his estate and onto another and then they might just pop over or under a gate, join up with somebody else's pheasants and be lost to his shoot forever.

We try to help each other, the Keeper and I, and many times I get a text at 4am, 5am in the morning, 'Cow calved, both OK.' So I for my part will try, when confronted by a group of his pheasants, to drive them back to the fields where they should be. But this is only half OK. If they are straying away, you can blow the horn at them and they will scuttle back along the lane and head homewards through a gateway. But if you come up behind them, 'going away', you have to get off the tractor and try to get in front of them and drive them back. This can easily turn into a 20 minute job, you don't always need 20 minute jobs when you are busy.

Anyway, I'm going along the lane one morning and come across 30 or 40 poults that are three fields off their estate, so I slow down and drive them back to where they should be with a bit of banging on the truck door and some horn blowing. Suddenly, without any warning, the whole bunch of pheasants drop down flat on the road. I look up, and there is a buzzard flying above them. I find this fascinating. These poults have been reared as intensively as any chickens until now but they still retain these strong natural instincts. I've tried to come up with some analogy of how they all dropped down together and unfortunately the only one I can come up with is a game we used to play on rugby tours called 'dead ants'. Someone would shout 'dead ants' and you all had to drop down on the floor, lie on your back with your arms and legs in the air. If you were last to achieve the dead ant position, you had to pay some unmentionable forfeit. We did it anywhere: in pubs, clubs, anywhere. If we were well ahead in a match we would do it then. I called it on one memorable occasion when we were all crossing a busy road on a zebra crossing on the Isle of Wight; it was memorable because the car that stopped to let us across was a police car. Not everyone takes the care that I do with these wayward pheasant poults. I have seen vehicles drive straight through them without slackening their speed, leaving a veritable carnage of dead and broken birds in their wake.

**SEPTEMBER 21ST 2013**

On the TB front, things are not as bad as I first feared. I had feared that we would be overrun with beef calves that we couldn't sell. Turns out that if we keep these calves in an isolated building, they can be sold in a TB-restricted market to someone who in turn has a building that is also suitably isolated. The fact that our local auctioneers now hold a monthly market for restricted cattle that attracts large numbers of both cattle and buyers tells its own story. It shows that the numbers of cattle that now come into this category are a significant number of the total and that there has to be an outlet for them. I didn't know all this because I quite simply didn't need to, but now I have to know because I'm right in there with the problem part of it. The auctioneer says the calves will make £50 less than 'normal' calves, but I suspect it will be a lot worse than that. I know that I will lose a lot of my calf income. The only plus in there amongst all the negatives is that we won't be having to slaughter young healthy animals, which was the greatest fear of all.

\*\*\*

It's my wife's birthday and we are out for a Chinese meal: my children and my grandchildren, all of our immediate family. The two youngest grandchildren still believe everything I tell them, stuff like the small scar on the back of my hand was caused by a machine gun bullet in the First World War. The two oldest smile indulgently; they know it isn't many years since they believed it as well. My efforts with them are now designed to teach them to give everything their best shot, to be cynical and not to take life too seriously, based on the theory that has stood me well all my life: don't believe anything you hear and only half of what you see. I've got a twelve-year-old granddaughter that sits in the middle of this age range who was chased by my stag turkey recently and demonstrated a vocabulary I didn't know she possessed.

My car was on empty, very empty, on the way there so I stopped to fill it up with petrol: I still can't get over how much that costs these days. We have a really good meal and even with very healthy appetites, there's still some left. We get the bill and I say 'I'll pay.' This should have been a cue for, 'No don't do that, we'll split it three ways.' Not a word, yet it's me that's on a pension! I mentally add the cost of the meal to the cost of the petrol and the only plus I can see so far is that I still have the petrol. But I get pleasure from my family, a lot of pride, and as I look at them as they chat away in the Chinese, I can't help thinking that my life is nicely complete.

Thinking on that theme, and developing it, the only thing that is missing, that has been on my mind for a little while now, is a pig. There's untold pleasure to be had from the ownership of a pig. If it's quality time you need in your life, there is nothing to compare with the precious minutes you can spend leaning on a sty wall, communicating with a pig. If everyone had a pig that they could talk to I'm sure that there would be a lot less trouble in the world. My brother's personality is much more rounded since he has taken to fattening three pigs a year in the wood at the end of his garden.

The first thing you need to contain a pig is a concrete floor: anything less and it will dig itself out. That's a given. They will get out in no time at all, they will make *The Great Escape* look like a long drawn out saga. Next a place that's nice and dry and warm to sleep in. So I get my laptop out and look at pig arks. They are around £500-£600 but that doesn't matter because you won't make any money out of a pig anyway so you can forget that altogether. The internet is OK for research, but for real information, you need to get down to the pub.

And that's where I go when I get back from the Chinese. There's a farmer there that keeps a few pigs. I used to take the grandchildren to see them and teach them to talk to pigs. He used to have Tamworths and Gloucester Old Spots and lots of crosses, so he had black pigs with ginger spots and ginger pigs with black spots and some pigs with stripes! He went into pigs about ten years ago. He bought about ten pig arks, put them on a concrete slab and put robust gates and partitions between them, a proper job.

So we discuss pig arks and concrete then he tells me that he doesn't have any at the moment, pigs that is. 'Why not?' Seems that they have totally destroyed his original investment, all he has left is the concrete slab. He has nothing else left of his original outlay that would keep a pig in. But he won't be beaten on this, he's going to build something more robust, a proper shed with concrete walls and steel doors. He only wants to keep about ten sows but he needs to be in control of them. And isn't this exactly what I was telling you recently, how pigs made the journey from pig sty to sow stall, because of their destructive nature? There may be those of you that will say that it is the hand of man that took the pig down this route by seeking to confine it. That would be true, sort of. But if you want to see real destruction, just let some pigs roam free. I'm determined to get a pig, just to address the challenge of its housing. As I drive home I mentally add the cost of the round I've bought to what I've spent already. You wouldn't want to know.

**SEPTEMBER 28TH 2013**

Next Friday I take about ten calves to market. This is good because at one time I thought the best-case scenario for selling calves would be four months. But they will be TB-restricted calves going into a TB-restricted market. Whilst I will be glad to see them go, because it will free up precious space, I don't expect to get much money for them. The buyers that are there know they will be cheap and will gather like vultures. The only plus amongst all the negatives will be that I haven't had to slaughter them. Usually these calves are an important part of our income, all I will be doing now will be getting rid of them.

*** 

We've got this big field. It's our biggest field, it's 45 acres. Well it's big to me anyway because it's only five acres less than the whole farm I started with. So it's big and it's sort of squarish and it slopes two ways and no matter where you stand, you cannot see the hedge on the opposite side. There's a frequently-used public footpath crosses through the middle of it and if the ground is bare and walkers cannot see the tracks of previous walkers, they often lose their sense of direction and get lost in the middle, end up at the wrong end, and have to walk the boundary until they find their next stile. In a way it has several horizons, which is important information for you as this particular story unfolds.

This year we have 11 acres of kale growing on it, (it was supposed to be 10, but there you go), and the rest is in stubble turnips. The kale is about as high as where I used to have a waist and the turnips are just over the tops of your wellies. It's the time of year when we have quite a lot of dry cows, cows not milking before they have their next calf. It's an important time for these cows, they build up their bodily reserves, get their feet off the concrete, it's a sort of bovine chill-out time.

So we take cows home to calve and we dry cows off. So there's a constant ebb and flow of cows in and out of this field, but there's mostly 35-40 cows there. They eat the turnips behind an electric fence, they can have as much as they will eat but we need to have some control lest they waste them and we don't want them to eat the kale yet because that's for the winter. So the electric fencing takes a bit of ingenuity because the cows like to run back to an adjoining grass field and the fences have to allow them access to the water trough. At the same time you have to avoid putting the electric fence across the footpath, for fairly obvious reasons. I'm trying to set the scene here: it ends up with two electric fences about 100 yards apart with cows and walkers down the middle between the two. Now cows will confound you and sometimes calve in the turnip field, that's no big deal, you usually just push the calf gently into the trailer and the mum will follow in behind. But calves like to confound you as well. If born outside they will sometimes secrete themselves in some long grass or in a clump of nettles and be desperately difficult to find. It's a bit like photos you see of baby deer curled up in grass, it must be some primeval instinct in them that is aimed at keeping them safe from wolves and the like.

Years ago, a calf went missing out of a pen so we assumed it had got out and hidden itself in this way, in the nettles around the sheds or in the bottom of a hedgerow. We looked for it, on and off, for four days. With the passing of each day our searches became more urgent until we thought the calf would be looking for us because it would be hungry. On the fourth day a Land Rover trundles up the yard driven by a friend of mine: he keeps beef suckler cows. He says he had a cow that had had a dead calf, he was desperate to foster one onto her, came on our yard, there was no one about, so he helped himself to one, how much did he owe us? Didn't mind him taking the calf, but a phone call would have helped.

Anyway to return to my original story, this cow calves in the turnips and we note that cow and calf are both fine so we go to fetch them home next morning and the cow is still fine but there's no sign of the calf. If you remember, the calf has 45 acres to hide itself in. Dilemma: the cow needs milking, so we take her home and go about four times a day to catch the calf. Sometimes we see it, chase it, but it disappears over the horizon, but by the time we get over the horizon, it's flopped down and we can't find it. Sometimes we go and there's no sign of it. Meantime it is doing fine, some of the cows in the field will be starting to get ready to calve, there will be some milk in their udders, they will be getting hormonal, they don't mind suckling a passing calf. As any dairy farmer will tell you, this is not ideal, but we are doing our best. So the calf is flourishing, getting bigger and fatter by the day. It is five days before we catch it, and this involves an intrepid leap out of a moving truck. It will go to market at the end of next week: let someone else chase it.

## October 5th 2013

There's been a lot of stuff in the media, nature programmes and the like, about the reduction in numbers of barn owls. I find this quite strange because there are more barn owls about around here than there have been for years. I thought that the increase here was down to the number of fields that had six metre margins around them to provide an area of rough grass where voles and mice could thrive. Voles and mice equalling food for owls. I've been saying for some time now that these margins didn't work because the rules on managing them were very strict and that allowed saplings to spread out from the hedgerows so that after a few years you didn't have a margin of rough grass, you had an elongated thicket that was of no use at all to the hunting owl.

I've been told by my landlord to plough up my six metre margins because they are full of this woody growth and ragwort. The main problem for your hunting owl is that the other place where you get this excellent rodent habitat is on verges of roads, and your said owl hunts at car height with the inevitable results. Off-hand I can't think of a more beautiful bird, its plumage detail is one of nature's greatest works of art. Unfortunately you rarely get the chance to see the intricacies at close hand unless the bird is dead on the road or in a glass case. Anyway, if the numbers are down, it's no good doing nothing about it, no good sitting on your hands, so I've bought an owl box and I'm going to put it up in an oak tree in a field next to the road where we often see an owl hunting. Well, Stephen will put it up in the tree with the loader. It's a simple, relatively inexpensive exercise that will probably make a difference. Far better than the millions of tax-payers' money spent on six metre margins that were of only short-term benefit. Come on, you other farmers, get out and buy an owl box.

\*\*\*

We had some farming friends up for the day recently. They wanted to have a look around the farm so I took them for a tour in the truck. A tour in the truck involves riding up tracks and at this time of year it involves the inevitable platoon of pheasant poults going up the track in front of you. They don't see pheasants where my friends live, so it was, to them, something of a novelty to see all these pheasants, especially for their children, so I'm driving very slowly so they can have a good look.

There are about 20 poults in front of us and they are all growing well now. They should be. Besides what the Keeper provides for them they can indulge in Nature's rich autumn bounty – shed corn on stubbles and plenty of insects. There have been lots of daddy-long-legs about lately and pheasants love them. I was doing some fencing the other day and paused to watch a group of pheasants chasing daddy-long-legs in the grass. It was almost like watching a group of spaniels searching for game.

Anyway these pheasant poults are going up the track in front of us when suddenly a buzzard comes over the hedge and grabs one of them. It takes off again, but with difficulty, because the poult is too heavy. At about ten feet high, struggling. It has to drop the poult back down onto the track. The poult lies there motionless, it looks very dead. The buzzard settles on the grass on the adjoining field but only about ten yards away. It seems to glower at us whilst it waits for us to drive on so it can return to its meal. For all of us, but especially for the children, it was a piece of wildlife drama that had unfolded in front of us.

***

Last Friday I took seven TB-restricted calves to the TB-restricted market. There were four black and white dairy bulls, pre-TB I had a private buyer for these who would pay me between £80 and £100 depending on the calf. In the auction these calves averaged £28. The other three were British Blues that averaged just over £120 total. I reckon they all averaged £50 a calf less than they would have. I've had to ask my bank manager for extra money to tide me over this crisis. 'Of course you can have some extra help, I was brought up on a farm, I understand what it's like.' Then they hit me with what they call an arrangement fee for this extra money, which means that it will take the money from these seven calves and probably ten more to pay it. Why does it feel as if we are going backwards?

## OCTOBER 12TH 2013

It's dark now when I drive to the pub. I've started going a bit later. I drink mostly an elderflower cordial. Elderflower cordial is good for you, it creates exercise. If I go a bit later to the pub I have one or two fewer elderflower cordials which means less exercise. I do not actually need to be walking about in the middle of the night! One of my other bad habits is putting my seat belt on after I've got the car moving. So I've gone about 40 yards, very slowly, because I'm sitting on the buckle, and I stamp on the brakes, I've seen something move in front of me. It's a hedgehog!

I can't remember the last time I saw a hedgehog on our yard. I've seen one on the road, dead, so far this year. Twenty or thirty years ago, a dead hedgehog on the road was a common sight. At about the same time as my encounter with the hedgehog, there is an article in our local paper predicting the extinction of hedgehogs by 2015, stating that there has been a 95% decline in the number of hedgehogs in our country. 'The brown hare is also endangered!' (I had a ride on the combine in our spring barley yesterday and in an hour ten hares came out). There have never been more brown hares about here. 'The decline in hedgehogs, hares and lapwings is all down to intensive farming.' No surprises there then! But hares, hedgehogs and lapwings all have one thing in common: their young all live on the ground. The air is full of predators by day and roamed by predators at night. One of the night-time roamers has a particular liking for hedgehogs, the badger. The badger is particularly adept at rolling a hedgehog over and eating it. So if there are massive numbers more badgers about, doesn't it stand to reason that they are all eating massive numbers of hedgehogs, or is that too obvious an explanation?

\*\*\*

My visit to the TB-restricted market has left a lasting impression on me. Normal markets are all hustle and bustle, all animation, there's a spring in people's step, people are striding about with purpose, there's an undercurrent of excitement. Not so with your TB-restricted market. True, it's a smaller market than normal, but the people about are subdued. There's a printed catalogue of entries, so very publicly, you are saying that your farm is affected by TB. There's a feeling of stigma about it – 'Look at me, I've got leprosy,' but none of us has done anything that puts us in this situation. I talk to other farmers who are more experienced than me, because they've been affected longer, and they tell me that the bigger and older cattle that are for sale, the nearer their value gets to the price of unrestricted cattle.

It's the younger stuff, calves and the like, that take the biggest hit. Just to endorse this view, some bunches of black and white weanling cattle go through the ring. They weigh 150 kilos. In a normal market I would guess that even on a bad day they would be worth £300 apiece. They sell for £75 each. A clear loss on what it will have cost to rear them. They were probably only being sold because the vendor needed the space they were taking up, and the money. The vendor probably went home heartbroken. Heartbreak and financial difficulties don't register on any TB statistics that I have ever seen.

\*\*\*

I suppose I should tell you about my honour. Get it out there in the open. Last Monday I became a bingo caller! There, what do you think of that? How many of you can say that? It just shows what a high calibre author Merlin Unwin Books gets. It's very serious stuff, bingo and older ladies take it very, very seriously. I was sure that I would get a bollocking but I came through unscathed.

## OCTOBER 19TH 2013

When I was a little boy there was a TB hospital, in a former stately home, outside our village. They probably called it a sanatorium, but I probably couldn't spell that when I was a little boy. Sometimes, on nice sunny days as we rode our bikes around the lanes, we would meet the patients out taking the sun and the fresh air. They were mostly people from the south Wales valleys and they all had an unhealthy pallor about them, which may seem a bit obvious, but it's an impression that lives with me still. They were ill, they looked ill and even then I suspected that they were the better ones. At this time I used to help out a bit on a friend's farm, (it's where the farming bug bit me), and I well remember his Dad coming home from market one day and saying that he had just seen the highest-yielding cow in the county sold for very little money because she had TB.

It was a mammoth task taking the nation's cattle from a position where TB was endemic to where it was TB-free. And like with all mammoth tasks, there were steps along that road to achieve this. You couldn't go from A to B in one big step. So whilst cattle were achieving TB attested status, it was perfectly legal for non-attested cattle to be traded. The irony in this particular story is that the high-yielding cow was sold to a producer retailer (someone who milked his own cows then sold the milk on his own milk round) up in the valleys. Which is where, if you remember, most of the patients in the sanatorium came from! Even as that little boy sitting at a kitchen table, I remember thinking it was a crazy scenario. The nation's cattle became TB-free in the 1960s. This was a huge step forward for the health and well-being of the nation and the nation's cattle. So why am I dragging all this history up? This week saw us undergo a 60-day TB test, which is the prescribed test interval after a breakdown. Five more cattle have reacted to the test. Ever the optimist, I was naturally hoping for a clear test.

Now things have gone from bad to worse. For nearly 50 years we have drunk unpasteurised milk straight from the cow, today we stopped doing that and bought some milk from the shop. (You can't get TB from pasteurised milk). To my mind, just having to do that is a step back down the road to that awful scenario, all those years ago. Is it any wonder that farmers are frustrated by what is going on, or more accurately, what isn't going on? We're used to adversity, it's a part of our lives, we can fight the weather and make the best of it. With TB there is nothing at all I can do. Fighting back usually makes you feel better: doing nothing and not being able to do anything just makes you feel worse. The good reasons that existed 50 or 60 years ago that made TB-free status a worthwhile exercise are still as pertinent today as they were then. That status was precious. Anything that is precious needs our care. If it doesn't get that care, see what happens. If any of you want to watch my five cows go for slaughter, you can. I have to.

\*\*\*

The last of our harvesting for this year involved baling our spring barley straw. It had a couple of showers on it after it was combined so I turned the straw swathes over just to dry them out. I'd had a ride on my nephew's combine when we cut this barley and there were a lot of hares in there. So I'm going quietly down the field, moving the straw over, and I'm actually marvelling at a red kite as it, seemingly without effort, covers the field looking for food. A slight movement of a wing here, an adjustment of tail feathers there. Suddenly a half-grown leveret breaks cover in front of me and makes for the next swath. Just as suddenly there are now three kites, two must have been behind me. The leveret makes it to the next swath but the three birds soon dig it out and it is carried off by the victor. As the kite goes past the tractor with the leveret in its talons, I can see the poor creature looking about at this new perspective on life.

## OCTOBER 26TH 2013

At this time of year, the most common bird around here, by a long, long way, is the pheasant. Some estates release birds in their thousands, some in their tens of thousands. We have 80 acres here at home, on which we have the shooting, where there are plenty of pheasants that stray here from various adjacent shoots and estates. We don't actually shoot here ourselves but we allow the next-door shoot to walk our ground. They get a couple of drives out of it. What I do like about this particular shoot is that a lot of my local friends from the pub are involved as beaters, do a bit of gamekeeping, and they get invited as a recompense, instead of rent, for use of my ground. Stephen, who works here, gets a couple of 'outside' days with his gun.

What I think is nice is that all of them, guns and beaters, mix together well socially, eating and drinking together. There is an added bonus to all this. When we are in the pub I can enquire of a third party if he knows where I can obtain the services of a part-time gamekeeper. As there are two or three in the audience who are involved with shooting the ground, and therefore think that they look after it, their faces are a picture. I go on to say that we are overrun with foxes, magpies and carrion crows, and that soon there will not be any pheasants about. It's a nice bit of winding up and is usually as stimulus for much gamekeeping activity. I've reported here, often enough, that we've been overrun with foxes this autumn and in the pub I've suggested a fine of £1 for every fox we see. I don't have to expand this idea further because there is always an audience in the pub that will jump on the opportunity to do some mickey-taking.

The fox bit is very true. My son David goes around the cows that are due to calve every night at midnight and there is a fox or foxes in the calving shed every time he goes there. The attraction for the foxes is the food they get from the discarded placentas. When the cows are fetched for milking in the early morning, foxes are always seen. One morning was particularly memorable. I will tell it in David's words. 'I was walking down the track to fetch the cows and my torch battery was low so I was walking in the dark. Mert was with me and I'd sent him on ahead to go around the cows. I became aware of something walking at my side and thought it was Mert, not doing his work. I scolded him, (at four o'clock in the morning, a scolding is actually a bollocking). Nothing happened so I switched the torch on. It wasn't a dog walking at my side, it was a fox!'

\*\*\*

Over the years I have written many stories of my dog Mert and my battles with the cockerel Neville. Mert is alive and very well, but he's getting on a bit now. I don't, anymore, take him for long days on the tractor. It's hard work for him to get comfortable, but he still goes everywhere with me in the truck. He still amazes me with how he can handle cattle. We were doing something one day with the dry cows; there were 47 in the group and dairy farmers will know what I mean when I say they were well full of themselves. Stephen had put a round feeder in their field whilst I drove around them to see if they were OK.

But Stephen left the gate open whilst he went to fetch a big bale of hay. They spotted this from 400 yards away and they were away. I tried to cut them off in their headlong charge by trying to overtake them in the truck. I managed to stop about half of them but those I didn't, 20-odd, were away down the track, bucking and kicking along. I got Mert out of the truck and said 'Get by'. Loose cattle around these buildings have multiple choices where to go, but they always go just one way. It's a magnet to them, a magnet called the landlord's garden. They went out of sight around the corner, we're talking here of over 12 tons of bovine rampage, with attitude. They were only out of sight a second and next thing I knew they were on their way back at some speed with Mert behind them, giving anything that loitered a sharp nip. How do you value something like that?

***

It's a different story on the poultry front. Neville has been gone inside a fox for about 18 months now. Occasionally visitors still get warily out of their cars and ask 'Is that cockerel still about?' I answer by saying that I haven't seen him for a couple of hours, which is perfectly true but obviously not the whole truth. With that information the visitor's business is conducted with frequent looks over their shoulders. But the Lord taketh away and he also giveth. So you lose the aggression of Neville on one hand and you get a stag turkey to replace him that is just as bad except bigger. Unusually for me, I haven't named him yet, I think it's because I was losing so many turkeys to the fox I didn't want to get too attached. I've only got three turkeys at present: stag, hen and youngster and they are confined to an old walled garden we have so that I can put them in a fox proof shed at night. The shed is about 20 yards from the entrance to the garden. If I want to go out and need to shut them up before it is dark, I have to fight the stag with a croquet mallet to drive him in. When I let them out in the mornings, I have to go as near to a sprint as I can now get, to try and get out. He always catches up with me before I get to the gate, so once again I have to ward him off with the mallet. Unfortunately for my wife, her clothes line is within this garden so she has had to take to drying the washing draped on her car. This is not a perfect solution because her car isn't washed very often and she parks it under the flight path of a colony of sparrows.

***

I was at a funeral, up in the hills, at a tiny church in a narrow little valley in a remote area. The church is supposed to hold 140 people and I wonder, as I wait for the service to begin, just how many people lived in these hills when it was built. There's a similarly-sized chapel in the next valley. There's hardly any room to park cars so most people at the funeral have had to park in a field half a mile away and I have no doubt that years ago, when it was built, most worshippers walked there. The area must have been teeming with families living in isolated cottages. This is sheep country and today the sheep farmers are out in force. They've had such a tough time this year that some will carry the psychological scars for a long time. One farmer near here is rumoured to have lost over 1,000 ewes in the spring, never mind the lambs! The family mourners arrive and the farmers who can't get inside crowd the porch to get out of the wind. The door is left open for them and I can detect the faintest smell of sheep and the slightly stronger smell of mothballs from their only suit.

## 2 NOVEMBER 2013

You will all be familiar with the maxim, 'Things can't get any worse', and then they do. Well I had my second excursion to the TB-restricted market last week. I took 16 calves, a mix of black and white dairy bulls and beef crosses. The whole lot averaged £29! Five of the dairy bull calves made less than £10. I'd kept and fed them all for three weeks or more, the ear tags alone cost £3. Other farmers there, who suffered a similar fate, said that they wouldn't go through the same humiliation again and would take any future calves to the hunt kennels. The auctioneer told me that he was now getting more calves at the TB-restricted market than he was getting at the normal one. We'll have to take stock of what we do from now on. If I told you what I was truly thinking as I drove home, I would probably get my farm burnt down. When I got home I worked out that on just that one load of calves, TB had cost me £1,000. There's nothing that I can do that will prevent the same thing happening again next month or next year.

***

We are out, Stephen and I, on our very highest field. It just touches the 980 feet mark. To enhance my stories of this high land, I always say it is at 1,000ft, which for me, is not exceptional exaggeration.

The field has been, for twelve months, in a fallow stubble, left for the wild birds. But it grows a crop of weeds and I am chopping these off and Stephen is coming along behind me with the plough. There have always been skylarks up here though there are fewer now than there were ten years ago, despite all the land that has been left in fallow for their benefit. This is an excellent cue for me to go on to my hobby horse about predators but I will move on.

The skylarks flutter about as I approach on my tractor: some of them do their fluttering-along-the-ground routine, which is designed to lead a dangerous presence away, but they are not in any danger from me. I soon become aware that there are three tiny leverets in the stubble. When I say tiny, I mean that I could easily encompass them as little round balls of fluff in my hands. As I work across the field, they scuttle about 20 yards at a time, further down the field to safety. They are heading towards the hedge; there's a field beyond that with plenty of grass, there's food there and safety. But they get to the hedge and then turn back. They are not together, you can barely see them, but individually you spot them making these short darts back up the field.

I phone Stephen to look out for them and he reports later that he saw all three go past him and onto the ploughing. On the ploughing lurks danger: it's where the buzzards and kites are eating worms and grubs: there's twenty of them altogether today. They soon see the leverets and that is the end of that. My fault of course, for ploughing the field. It makes me realise just how little I understand nature. Here we are mid-October, this big field is very exposed to wind and weather, a most inhospitable place in winter. If I hadn't disturbed the leverets, how would they have survived? What on earth do the skylarks live on?

*** 

One of the wonders of nature that fascinate me are acorns. As I write, we are in the middle of the acorn harvest. After a windy night, the road under an oak tree is covered with acorns and the acorns themselves are covered with pheasants gorging on them.

What intrigues me is not just that pheasants manage to swallow the acorns whole, because in a previous life I used to handle thousands of pigeons and in the autumn their crops would be full of acorns, which for a pigeon is a much bigger beak-full than it is for a pheasant. No, what intrigues me is how they digest them. That hard shiny shell must take a lot of grinding up.

Acorns can catch you out. Cattle like acorns. In normal circumstances they will eat them as they fall and no harm done. But, without thinking, you can move them to a fresh field where there may be an accumulation of acorns and then you can get sick cattle and even deaths.

***

I went to the doctors yesterday. I went to get a flu jab and to give blood samples for my MOT which lapsed some years ago. What a sight in the waiting room! Packed out. Someone once told me that doctors can predict very accurately who will be in to see them on Monday mornings because 90% of them are there every Monday morning. I came out feeling very healthy and well, and I'd not seen sight of a doctor. Everything in life is relative and compared with what I saw in that waiting room, well and healthy is what I am.

### November 9th 2013

Not many years after I started farming on my own account, I was still driven by trying to milk more and more cows, albeit on a small acreage. I was getting my numbers up into the 80s and decided that my butterfat was a bit low. The quick fix for getting butterfat up was and still is, to buy some Jerseys.

So I went to a pedigree Jersey dispersal sale in the Midlands with the view to buying some older cows that were in milk but also back in calf again. It is often traditional at auction to sell the oldest cow in the herd first, and so on until you get to the young stock. I bought the first cow into the ring for £33, way below what I had expected to pay, and in no time at all I had bought nine cows for not a lot of money. As the auctioneer moved on through the catalogue, towards the younger cows, he moved out of my price range. Then he came to one of the highest-yielding cows in the herd. There were two notable things about this cow: she had given over 1,000 gallons (remember gallons?) in her previous lactation, which in those days was a lot of milk for a Jersey, and she was one of those Jerseys that you see occasionally in that her coat was so dark it was almost black. I'd had a good day thus far, bought more cows for less money than I had anticipated, so I decided to buy this black Jersey, which I did, for £100. Within the context of the day and within the context of what I had bought thus far, that was a lot of money. I had actually bought the dearest cow and the cheapest on the same day. I ended up with ten cows for £500.

The nine older cows lasted me a couple of years and when their productive lives came to an end I sold them on for very much what I had paid for them.

But the black Jersey seemed to go on for ever and ever, and what a good cow she was. There came a time when she didn't walk very well. She had a problem with her pelvis: she would go a couple of steps and then she would stop, put a back leg out as far as she could, shake it about a bit, there would be a loud click in her pelvic area and off she would go again. Apart from that, she was fit and well so we put her in our walled garden to live, and allowed her to suckle calves. We would put a calf with her, a beef cross bull calf, and after a month or so of love and affection, and a bellyful of Jersey milk every day, it would be a tremendous calf. So we would take that calf to market and put another calf with her. She would greet the new calf with all the kindness that she had shown to the previous one and so it went on.

After a couple of years of this, her milk dried up and it was her time to go. There used to be a small local abattoir ten miles away (it's not there anymore, which is called progress), she went off in a nice clean little lorry and ended her life about half an hour later in as kindly a fashion as possible. I know she had a good life here, and she was good to me in return. So what's with all this reminiscing? Well there's nothing new in farming. We get paid for the litres we produce and the butterfat content of that milk. So to try to enhance the value of the litre we sell, once again I turned to Jerseys.

Just over 12 months ago I bought ten maiden Jersey heifers that I could calve down this autumn. I had to pick the ten out of a bunch of over 20. Amongst this bunch was a black Jersey that was the image of the one I've been telling you about. I decided immediately I would have her but I was nonchalant about it. I picked her out about fifth. 'And I'll take the black one.' Move on 12 months and the Jerseys are calving down. The black one calves OK. She looks really well, nice udder, plenty of milk, a new favourite. Seven days after she had calved and with her whole productive life in front of her, she fails the TB test and is gone on a lorry.

*** 

The local shoots had their first day last Saturday so the conversations in the pub move inevitably on to shooting. One day out is not necessarily very fruitful for anecdotes so we have to listen to reminiscences of days gone by. We're having some of these stories for the umpteenth time, and it's amazing how they get embellished with the passage of time, not something I would do myself, naturally. So I tell them the story of when we shot a fox one Saturday and someone took it in the boot of his car and placed it with its head peering out of a fence at a shoot the following Wednesday. At the appropriate drive we were told there was a fox in this spinney killing pheasants and to look out for it. A certain gun was put in the right place and told to keep a look-out. Not many minutes later a cry goes out 'There it is'. There are two shots in quick succession. 'I think I've got it!' We can now see the gun trying to run laboriously across some heavy ploughing, letting off two more shots as he went.

I won't tell you what he said when he got to the fox but he has a very loud voice and most of the neighbourhood heard him.

## NOVEMBER 16TH 2013

My little granddaughter has wanted a puppy for a long time. It is of interest at this point to consider what she probably had in mind. It would probably be small and cuddly, obedient, probably wouldn't mind being dressed up as a Barbie, probably wouldn't mind being pushed around in a pram, you get the picture. What did she get? A 12-week-old sheep dog. Whisper it quietly, but this is part of our succession policy. Well, Mert is getting on a bit now and rather than wait until he can't, you get a successor in the pipeline. And if you need to do that anyway, why not tell the little girl it's her puppy? It's a win-win for everyone, except her. She managed to fuss it for a couple of weeks, but it soon got too big and strong for that sort of nonsense. I felt a bit guilty, but not for long.

So now it spends its days on the farm. It's called Maddy – at least we let her name it! It's a bit headstrong, which is another word for wilful: it does its own thing. Mert completely ignores it unless it gets within five yards, after which he attacks it. The turkeys are now in more danger from Maddy than they are from the fox, so for them it's danger night and day. She's different to our other dogs. If I go into a shed or the house, Mert just stands there waiting for me to come out again. This new dog takes one look at the closed door, assumes she is dismissed for the time being, and is off to have a look around. It's not just a casual look around, it's done at some speed until something takes her interest. There was nothing in her CV to suggest that she was interested in shooting, but the lads who shoot here came for what they call an outside day, so, unknown to us, she went off with them for a day's shooting and they brought her back when they had finished. Apparently she was quite good at finding pheasants. Which is a bit of a plus, because all she seems to find on the yard is trouble.

\*\*\*

My two eldest grandsons now attend a sixth form college 15 miles from here and probably 25 miles from where they live. Several local children, or are they students, make similar journeys and as some of them, including my eldest grandson, now have cars, getting there in the mornings seems never to be a problem. But they all seem to finish at different times and getting home is not so easy. Number two grandson seems to finish later than the others and has to be helped more. Public transport will get him half way home but we often have to fetch him the last half because his Dad is still working. After having to fetch him all the way one night, because he hadn't any money on him, I gave him £20 and told him that it was to spend on a bus or train and not on cider. I'm not sure it should be necessary to tell a 16-year-old not to buy cider but there you go.

Last night he phoned up at 6.30 to say he was stuck the seven miles away so I fetched him. I get him in the car and say he is late, how long has he been hanging about there. Turns out he's been there two hours. 'Why didn't you phone earlier?' Didn't have any more money, so he wandered about various people's houses he knew to see if they would phone us, but they were all out. In the end he went to a filling station and they let him use their phone. 'Where's the money I gave you?' 'I left it at home.' 'Where's your mobile phone?' 'I left it at home as well.' 'You could do with a spinner every morning.' 'What's a spinner?' And so, finally, we get to my story.

I once worked for a month on a farm in North Wales. I was there to help with the milking. The boss and I used to milk together in the morning and I milked on my own in the afternoon. There was a local lad from the village worked there as well, doing general duties. We milked, in those days, in a large cowshed and every morning there was a ritual to be followed. We'd usually been milking an hour when the lad from the village came to work and every morning he came and stood by us as we milked. Sometimes the boss would leave him standing there for ten minutes, just waiting. Eventually he would say 'Take your cap off Arthur.' 'Oh, do I have to?' 'Just take your cap off.' That was the ritual. Every morning Arthur had to have a 'spinner', to make sure he was fully awake. And what a spinner it was! The boss was in his early 30s, strong and as fit as anyone I knew. He had hands like shovels and fingers like pickaxe handles. He would make his hand into a fist with the knuckle of his fore finger most prominent. Although Arthur knew it was coming soon, he would always catch him out with a stinging swing of his knuckle to the back of the head. And Arthur would say, 'Ouch' rub his head where the hurt was, put his cap back on and go off to do his work. He was certainly fully awake! He wasn't that sharp, that Arthur, or he wouldn't have stood there like that every morning waiting for his spinner. It isn't a sad story because apart from making sure he was fully awake, the family there showed him lots of kindness for the rest of the day. I explained to my grandson about a spinner as we drove home. 'If you go to college again without any money and without your phone, you'll get a spinner.' 'OK' he says. But we both know he won't.

### November 23rd 2013

I like working with cattle, especially our younger cattle. Today we are sorting out heifers to put with the bull for the first time. This is a piece of work that I particularly enjoy. We aim to calve our heifers at around two years of age with their first calf which means that they go to the bull at 15 months, which always seems no time at all since they were young calves. Today we get in a group of 30 heifers to sort out the ones we need. Selection is based on a simple principle: if they are big enough, they are old enough.

I'm expecting that about half will be fit for purpose, but when we get them into a shed they seem much bigger than they were in the field and we find we have a group to put with the bull of 24. We take these in three trailer loads to graze off two fields that are adjacent to the sheds where they will spend the winter. The next job, the obvious job, is to fetch the bull. We usually put our heifers to our Limousin bull who is a couple of miles away with the 20 or so heifers that are due to calve in the spring. This is the bit that I am especially looking forward to. The bull is not that easy to load on his own, so I decide to bring one heifer with him.

That's the special bit. Last year I happened to buy a pure-bred Brown Swiss heifer and we'll bring her. Pure Brown Swiss are big gentle cattle, usually a sort of grey fawn colour, with a grey circle of hair around their muzzle. When I was in college we had an animal husbandry lecturer who used to ask us, 'What three species of animal has a grey mealy ring around the muzzle, boy?' He was from Devon so he called everyone boy. And the answer? Jersey cows, Swaledale sheep and Exmoor ponies. Clearly he hadn't come across Brown Swiss.

If I use just a bit of imagination I can picture this heifer grazing an Alpine pasture amongst the sweet herbage and the flowers. Put a bit more imagination into the picture and I've got Julie Andrews and the *Sound of Music*. It's something of a standing joke in this house, that the *Sound of Music* is always on TV just before Christmas and I am watching it when I am called for my tea. I ignore the call, and two or three subsequent calls, so that I have to be fetched. 'Can't come yet, I have to see if that nice Von Trapp family gets away safely.' But there's still one thing missing from this imaginary idyllic scene and that's an Alpine bell. Fortunately I can help you there. Because hanging on a nail in what we laughingly call a workshop is just that, the Alpine bell sent to me by a reader of *West Country Life*, years ago. So we've turned the heifer out with her bell on and she looks a picture. We'll see how long it takes someone to complain about the noise.

***

'They' had one of those stop and search blitzes outside our local town recently. Big gangs of officials, checking everything from road tax to insurance, to vehicles, to fuel. Star of the 'show' was an artic lorry that had come from Kent which wasn't taxed, had no insurance and was on red diesel. Apparently 120 checks were made, 20 vehicles and trailers impounded and 50% were running on red diesel. Nice to know that anarchy is alive and well and living in the countryside.

## NOVEMBER 30TH 2013

Today I have some interesting snippets of information for you. And as Christmas is getting nearer by the day, it's a good time to have a turkey update. It hasn't been a very good year in the turkey department. Firstly we had the ravages of the fox which put an abrupt end to the concept of free-range turkeys that lived and slept where they pleased. There's always an irony in farming. The irony for me with the turkeys is that all through the bad cold weather of last winter they spent their days in the relative warmth of the cattle sheds and their nights on a rail on a fence in the yard, exposed to all the snow and cold that the elements could throw at them. I feared for their survival in that cold but they coped with it admirably. What they couldn't cope with was the fox that plucked them off the rail one by one.

When the stag went, leaving us with just two hens, we had to abandon the free living concept and put them into our walled garden where there is a shed to shut them at night. We bought a new stag and it is he that has replaced the cockerel Neville as the danger in my life. Another hen has died, (turkeys like dying, it's second nature to them) but the remaining hen proved to be a prolific egg layer. She laid eggs in the nettles in the walled garden to start with and these provided sustenance for the next generation of magpies and carrion crows, so we had to confine stag and hen to their shed. The hen provided about 20 eggs for me to put in a friend's incubator. Think seven hatched out but they were weakly and today I have one left and there is one with the incubator owner. We let the hen sit on the last ten eggs she laid. Seven of those hatched out but none survived more than 24 hours.

\*\*\*

I've been to the doctors three times recently. I've never been much and haven't had a proper check-up for years, so two of the visits were for routine blood tests (lest you should fear I am ailing). We once had a doctor that I used to play rugby with and that used to be more fun than a visit is now. When my son and I were both playing rugby we would take any injuries down to the vets to be x-rayed and then take the x-ray to the doctors. This system saved endless hours of sitting in A&E. My son was always getting cut about the head, probably from playing in the back row and popping his head up out of the rucks to see what was going on. I played in the front row and always kept my head down in the murky depths. He always went to see the vet to be stitched up, vets providing a much better service than doctors.

He once had a huge cut under his eye with something like ten stitches in it, the eye itself was swelled up like an orange. He went to the pub as normal on the Saturday night and the landlord eventually asked him to leave because the sight of his eye was upsetting his customers. The vet used to put the stitches in and I took them out, we must have saved the NHS a fortune.

When I was last at the doctors I knew most of the people in the waiting room, whose number included the local undertaker. This fascinated me and I had to make a comment on it. Was he there to give his business card to anyone who looked really under the weather? Was he there because the doctors has given him a tip off? Indeed, was it possible that the doctors had made him a sort of partner so that their own service was literally from the cradle to grave? He didn't reckon much to my questions and was quite relieved when he was called to see a doctor. Some of the other patients looked a bit uncomfortable as well.

***

I met a very decent hard working young man yesterday. He used to milk 80 cows and rear enough replacement heifers to maintain that herd size so he hadn't brought any cattle onto the farm for about 15 years. This sort of family farm is the backbone of the UK dairy industry. A few years ago, TB was detected on the farm and today he has just 20 cows and two in-calf heifers left. What sort of society do we live in that allows something like that to happen? I asked him when he would restock. 'Never. Why would I want to put myself and my family through all that again?'

## December 7th 2013

I went to see Wales play South Africa and amongst the pre-match build-up was a tribute to the late Cliff Morgan. I have memories of my own of the man. When I was a sort of youth, a gang of us used to go into Cardiff on Friday nights. Our favourite destinations were the pubs down Bute Street. Bute Street was the road that ran from close to the centre of Cardiff down to what used to be Tiger Bay and the docks. It was a very different world down there compared to the world that we lived in on the farm. We would mix with sailors from all over the world and stare open mouthed at the prostitutes plying their trade. Stare, I said. Sometimes we would hear sailors planning a day time burglary of some sort in Cardiff next day, secure in the knowledge that their ship would sail on the afternoon tide.

At the end of the evening, perhaps midnight, we would make our way back towards the city centre and visit a place called Frenchie's Steak Bar. It wasn't posh, probably a long way from that, and it was presided over by this huge Frenchman who was probably nearer 30 stone than 20. You could only have steak and chips and a big fresh tomato and all the steaks were cooked in one huge frying pan, possibly four feet across. Time changes perception, but even to a man of Frenchie's size, it was a two-hander. Most of us in there that late had had a night on the beer, some more the worse for wear than others. Cliff Morgan was often in there, sitting at a table at the back with two or three friends I always assumed were colleagues from the BBC, having a meal after working late on a programme. If you are a famous rugby player, there is no anonymity in south Wales and Cliff would soon be spotted.

Often someone would weave their way to his table, plonk themselves down and proceed to harangue him about rugby. For the rest of us, this invasion of his privacy was cringingly embarrassing, but Cliff for his part was so courteous, gave them his full attention, and never sought to bring the encounter to an end. Gracious was a word that used to come to mind. Even at that age, with all that beer inside me, I could see that there was a lesson there to be learnt.

\*\*\*

In the pub and a farmer asks us if we know a certain person. *Do they know him?* They give him the person's pedigree going back about four generations and do one or two sideways excursions on his family tree to where there was a particularly juicy bit of gossip. He resumes his story and tells us that he had been to buy a sheepdog pup off this person. 'He told me that his place was difficult to find so the best thing was to meet him in the pub in the village at two o'clock.' Apparently one thing led to another so it was six o'clock before they went to see the pup. 'We went into the kitchen and I looked up and there was a big hole in the ceiling, and there must have been a hole in the roof as well, because I could see the stars shining.' His audience was unimpressed. 'Oh, that hole's been there for there for years, it got so bad one hard winter that they bought one of those second-hand mobile homes and moved out of the house into that. Then that got a hole in the roof too so they moved back into the house.' They continue to move the conversation on and it is only me that is polite enough to enquire if he bought the puppy. He had, so I ask him if it's any good. 'It's a nice enough pup alright but it's taken my glasses out on the yard and I can't find them.' You don't get stories like that sitting at home watching television.

## DECEMBER 14TH 2013

We've always been very lucky around here with regard to security. Honesty and trust were always the order of the day. Up until ten years ago we didn't even take keys out of cars at night and we only stopped doing that because I discovered that my insurance company wouldn't pay out on a stolen car if the keys were in it! All that has changed now petty theft is widespread. The main item to be stolen is diesel. Diesel disappears from farm tanks, vehicles and tractors on a weekly basis. Sometimes the tractor, for example, can be parked miles from anywhere, completely hidden from view, so you wonder just who is about 'spotting' these opportunities. I know a farmer who had had his diesel tank emptied on a regular basis no matter how he tried to secure it. In the end he bought a new tank and hid it away from view in another building. For good measure and as an act of revenge, he filled the original tank with water. Within a week the water had gone so he put some more in. 'They' have never been back since. There is a story about a local man needing a new fuel pump and injectors in his truck. Shame that.

### DECEMBER 21st 2013

I've not yet decided how grumpy and miserable to be over Christmas. My disposition at present is quite sunny, for me, but I have a reputation to think of and it has taken many years to acquire it so I will not cast it aside lightly.

My first 'outside' job of the day is always to let the turkeys out. I like to let them out before it gets light because they stay in the shed whilst it is still dark and this reduces my chances of being attacked by the stag by 50%. By applying the same clever logic at night when they are safely in their shed, I don't get attacked then either. In fact if I didn't have to feed them I'd be quite safe every day. I do have to feed them but I can usually put out enough food and water for three days. I then go back into a nice warm kitchen and do some writing. Then I put my pen down and have another reflective cup of tea. It's just getting light. Outside our kitchen window is our bird table. The first visitors today are two turtle doves. This is a nice Christmassy omen, not such a good omen for some of the turkeys, two or three of which will lay down their lives for Christmas in the line of duty, but we might just as well eat them as the fox. Turtle doves interest me. I read that they are in such decline that they are endangered. I struggle with this. Farmers are to blame, as usual. Never mind that these doves have to undertake an arduous migration. Never mind that there are great changes to their habitat in Africa. Never mind that they are affected by some disease. Never mind that Europeans try to shoot them as they fly there and as they fly back. It's all down to farmers like me which is always the easy shot. We have a lot of turtle doves around the yard; quite when they migrate I'm not sure, it could be that they get much cheaper flights on Christmas Day, which I have heard is a very cheap day to travel. I know there were turtle doves here all winter last year so I just wonder how inhospitable farmers really are?

***

I can sit in the pub, quietly, and listen to what goes on, and a topic of conversation can start on quite a mundane level, and before you know it, you have a gem. It was a cold and wintry night, not many of us there, and someone commented on how good the fire was that was keeping us warm. And it was, just an old fashioned fireplace, nothing fancy, but throwing out lots of heat, but at the same time not burning buckets of coal and lots of logs. It was doing very nicely on a mix of these two. Then someone says that the fireplace didn't used to be where it is, it was once the other side of the wall, same chimney, different room. So then we have a recap of all the alterations that have gone on in the pub, over the years. The bar used to be there and before that it used to be there, all that sort of stuff, the detail of which, I can't personally remember. I do remember that the longest room in the pub, now the restaurant, used to be inaccessible from the pub, only from outside. This was because I think the pub had been built as part of a large estate and this was called the parish room, I can remember having to go there to pay the rent many years ago.

So then we have a recap of all the previous landlords in reverse chronological order. I can't contribute to this either. Then they start on characters and end up with two retired farm workers, now long gone, who were regulars in the pub. Before we address their story it is probably important to consider their background: probably born in the depression years of the 1920s, almost certainly sons of farm workers, which would mean very low wages and poor living conditions, though to be fair, probably no worse than the lot of most manual workers at that time. They would have been young men in the difficult years of the 1930s, then came the war years. Followed by the austerity of the post-war 50s. What is to be sure, they would have had harder lives than any of us can imagine, and that in itself would have a bearing on this story. Both of these men were in the habit of putting their false teeth into their beer glass if they had to leave their glass unattended whilst they went outside to the toilets, as a deterrent to anyone drinking their beer while they were away, which was its purpose, and it was very effective. But one night they both went out to the toilet at the same time and someone swapped the sets of teeth over. They didn't tell them what they'd done for two weeks and then it was only because everyone was fed up listening to them complaining their gums were sore.

## DECEMBER 28TH 2013

I had gout at the weekend. It was in the joint of my big toe, the joint that joins the toe to the foot. An important joint for walking on. Why does everyone think gout is hilarious? Is it because in our mind's eye we have a picture of a heavy drinker, with swollen nose to prove it, with his foot up on a stool, he has a huge swollen big toe with exclamation marks around it, to signify the throbbing that is going on? If you've ever had gout you will know that there is nothing funny about it at all. I went to bed early (Friday night is one of my early nights) with quite a pain in my toe. We'd been TB testing that day so I assumed I'd knocked it or a cow had trodden on it, though I couldn't remember either event. An hour later and I had to get up, I just couldn't bear the weight of the duvet on it. I spent the whole night watching the Second Test from Australia, which was worse than the pain in my toe. Because it is now Saturday and I need a prescription there are further problems. You can't phone our doctors at weekends, you have to phone an agency that covers the whole county. You end up speaking, eventually, to a doctor who doesn't know you, has very little idea of where you live. If you can't phone your doctor two days out of seven (which must be over a quarter of the time), is it any wonder that A&E departments at hospitals are overwhelmed? So I get to speak to a doctor. 'I've got gout.' 'How do you know it's gout?' 'Because I've had it before.' 'When was that?' 'About 20 years ago, I've still got the tablets left from then, is it OK to take some?' He gets quite animated now, concerned that I might take these old tablets. He says he will give me a prescription for some new ones.

Trouble is, he is 25 miles away. He is explaining where I will collect the prescription and I end up explaining to him that in this modern era it shouldn't be beyond the wit of man to email a prescription to a chemist which will only be six miles away. I don't actually use the word 'wit', just in case he ever gets close enough to me to stick a needle in, but you get the drift. So he says he will do that and that the prescription will be there in half an hour. We go, various of us, three times to get the prescription, and such was the speed of modern communications, it was there in six hours. It crosses my mind that the three trips are a fair proportion of the 26 miles we were faced with to begin with; it also crosses my mind that the doctor knows I didn't use the word wit, but that was what I meant. I don't want to go on and on about gout, but it hurt so much I didn't go to the pub on Saturday night, so you know how serious it was.

But I do get there on Sunday night, and after they have all had a go at me, still limping a bit, and asked how much port I drink (none), they notice that Stephen is limping as well. So then we have a discussion about gout being contagious! But Stephen is limping and because he sees me getting all the attention he goes to the doctors as well. He comes back quite pleased with himself and says he has policeman's heel! Never heard of it, have you? Must go back to the days when policemen put their thumbs firmly into their lapels, rocked back and fro from toe to heel, and said 'evening all'. It's the only explanation I can think of, we must be talking *Dixon of Dock Green* at the very least.

\*\*\*

It was an emergency. Would I call in at the village and inspect a turkey and assess its suitability or otherwise to be a Christmas dinner? It's the turkey I was telling you about that was conceived here and found its way as an egg to be incubated under a bantam hen. Its name, apparently, is Gilbert. I picked it up, ran my hand over its chest and said it would be fine, as long as they only wanted one turkey sandwich. As he belongs to a family that are clearly not given to sharing one sandwich, we decided that he would not make a Christmas turkey, well not this Christmas anyway.

I think the trouble was that he was living in quite a small pen with his 'mother', the bantam, and three ducks. He was clearly suffering from an identity crisis. He was probably using up all his energy trying to quack. So we've decided that he will come here so that he can mix with his own species (and his mum and dad), have a good run out on grass and make a nice Easter turkey instead. Hoping to get an invite to that particular meal. One of my favourite films is *Babe*. There's a duck in that which thought it was a cockerel and it seemed to cope with that fine.

## JANUARY 4TH 2014

Some of my best friends are turkeys. So I now have two fewer friends! I didn't enjoy going amongst my flock and selecting two for Christmas meals, but I did: that was the main reason I reared them, or was it the company? Anyway two laid down their lives, I comfort myself by thinking I saved them from the fox. I didn't enjoy feathering them, never been big on feathering, me, not sitting on my own in a freezing shed all morning anyway. So we will start the New Year with seven turkeys, and they will be joined at any time by Gilbert the thin turkey who lives in the village, who is destined to be an Easter turkey. Will that be enough turkeys to last for 12 months? Time will tell.

*** 

I've been thinking about holidays. My sister can't get far without a wheelchair now, so we make a bit of effort, as a family, to see that she gets a holiday. We've taken her abroad lately but it's not that easy. It does us all good to have a week tending the needs of a wheelchair user, so that we are aware of just how difficult life can be for them. I couldn't go last year but they ended up at a hotel abroad with half a mile of a sort of crazy paving in all directions that was almost impassable. We are looking for a nice holiday let in this country this year. Downstairs bedroom, harbour or river view, flat walks for the evening, not as easy as you would think.

This made me reflect on a farming family I knew in South Wales. They never went on holiday. But a sort of concession and acknowledgement to their year's hard work was to go to the beach for the evening twice a year. The beach was only half an hour away so it was not that big a concession. The two occasions were the completion of hay-making and later after the corn harvest. The hay-making was a lot of hard manual work, the corn less so as they only had a couple of fields. Anyway, off they would go with a picnic and for a swim. One year they went for their swim and put all their clothes on the rocks. When they were getting dressed after their swim, the old granddad reckoned someone had nicked his waistcoat. It hadn't been nicked at all, he found it when they went back a couple of months later, after the corn harvest. It was under his vest!

## JANUARY 11TH 2014

Largely split up now, probably by death duties, the area where I live was once blessed by some large estates. I use the word blessed quite deliberately because there are still wonderful legacies from those days. Chief amongst these, are as far as I am concerned anyway are some wonderful avenues of tree. The one I have in mind this morning stretches probably two miles along a busy B-road. When I came to live around here, which is nearly 50 years ago, there was a tree about every 30 yards on both sides of the road. I suppose there are about 40% of those left today, the rest having fallen victim to gales and the chain saw. Every time that I see the demise of a tree it is with sadness, but if the tree has gone down to the chainsaw, you can always see that the base of the trunk is rotten. The trees in this particular avenue are all Turkey oaks. Foresters tell me that they are of little value, that they don't even make very good firewood! Workers on what is left of the estate tell me that 14 more are earmarked to be felled, which is a big, but inevitable, shame. I wonder how much of this decimation of the avenue is due to the culture of society we live in today? If you owned a tree on the side of a road, would you put your hand up and say that it was perfectly safe? Of course you wouldn't. If you were going to err, it would be on the side of safety, because if the worst happened, and someone should be killed, the blame culture would be all over you and you could even be facing manslaughter charges! This culture of litigation pervades all of our lives. I wonder how many people would go to a busy A&E department if they had to get past a 20 stone matron, who was there to filter out malingerers? There's a cost to this culture, a cost that we all share. It's a cost driven by society in general and nurtured by people in yellow high-viz jackets. That's enough of that, I can feel that I'm getting off on one of my pet subjects.

If I go back to my starting point, avenues, my favourite local avenue is one of horse chestnuts. When they are in flower in the spring they are a glorious sight. Seen from a distance they resemble two rows of hydrangeas in bloom. Chainsaws have taken their toll here as well, not always with good reason. But they seem to be very prone , now that they are mature, to losing quite substantial boughs, which spoils their shape, although they still remain an enduring feature of our springs.

***

My florist friend does a lot of flower arranging demonstrations, she goes over quite a wide area doing it. I was sitting next to her at a dinner recently and complimented her on her shoes. 'Very important shoes, for flower arranging,' she says. So you have to ask why, don't you? 'Well,' she says, 'I'm mostly up on a stage and my feet are at the audience's eye level. So I put all these big boxes of flowers in my van. I drive 100 miles to the venue. I spend two hours arranging the flowers, explaining all the time what I am doing and why I am doing it. Then when I am finished I ask them if they have any questions. Do you know what they always ask first?' 'No idea.' 'Where did you get your bloody shoes from?'

## JANUARY 18TH 2014

Some of our thoughts will be with those affected by flooding and loss of power. The series of low pressure weather cycles that have battered and soaked the UK have been spectacular in their delivery and on weather forecasters' charts. A blocked drain once diverted an inch or so of water into our house years ago. We thought it to be no big deal and also thought we had made a good job of clearing it all up. The stench from the carpet after a couple of days was unbelievable. So if you sometimes think that flood victims are 'trying it on' for new furniture and compensation, well they are not, believe me. People around here, well those with a rain gauge, reckon we've had about five or six inches of rain in ten days or so. I tend to use more practical yardsticks. My 'flock' of turkeys usually drink a bucket of water every two days. I haven't had to put any water out for them for a week.

As I fight my way about the yard against wind and rain I can often hear gunshots. Gunshots are heard every day, except Sunday, around here. So when I spare thoughts for others, I often think of those unfortunates who have spent thousands of pounds to stand at the end of a wood in a howling gale. If the wind is blowing in the direction that the pheasants are meant to fly they will come out on the wind like missiles. If the wind is against them they will get up and execute the fastest U-turn you will ever see and return back over the wood whence they came.

I used to love going shooting but I never spent that sort of money on it, not only because I've never had that sort of money, but I hope that even if I did, I wouldn't want to spend it like that. I'll be really glad when this year's shooting season finishes. Most of my Saturday night company goes shooting or beating on Saturdays, a day that ends with a meal in the pub at about 4pm. They used to, in the past, stay there until about 7pm then go home to change and come out later. Whether it's the weather or the convivial company, I'm not sure, but this year they've been going home at eight to 8.30pm with hearty cries of 'see you later', but later never comes. They get home, shower, sit down by the fire and that is it, not seen again that day. I suspect that the drink and the fire closes their eyelids. Which leaves me without their company. I spent one Saturday night talking to a man who owns a laundrette about the intricacies of washing machines. To my credit I haven't shared the conversation with you.

**JANUARY 25TH 2014**

It's still January and already I'm looking for harbingers of spring. This possibly tells me two things, a) that I am wishing my life away and b) that I've already had enough of this winter. My favourite harbinger of spring, by some distance, is the daffodil. I just love to see the first daffodils appear. One of the current news topics this week is the need to buy food produced in this country, in season and not to assume that we can buy anything at any time we like because it is produced somewhere else in the world and can be transported here: there's a carbon cost to all this. I've already seen my first daffodils, in my florist friend's shop, grown in a greenhouse in Holland (at what carbon cost). I'd rather wait for February and March.

But my impatience for the spring is already being tested. There are pigeons living in the roof of a barn we have and they have young ones already. Some have fallen out of their nests and are living on the floor, luckily for them it's a cat-free building. We've brought some heifers inside this week and the sheep of a young friend have now gone. Both these fields have greened-up so the grass is growing.

And that other tell-tale sign is there. Cock pheasants are already fighting over territories and have started chasing the truck, and the shooting season is not over yet! What I do know about weather thus far this year, is that our land has never been so wet and you take the truck beyond a gateway at your peril.

\*\*\*

It's a Tuesday morning, about 9am, and a procession of three helicopters make their way slowly up the valley. They pause half a mile away, as only helicopters can pause, then one by one they drop down behind the tree-line to land. There's a shooting lodge there and the helicopters are delivering a load of shooters for a day out. That's the sort of big money that's involved in large commercial shooting. Mostly the 'guns' are delivered by a convoy of Range Rovers, but today it's a convoy of helicopters. Everyone else will have seen the helicopters and by tomorrow the rumours will abound as to who they were carrying. Someone will say it's Manchester United footballer squad, someone else will say it's the Rolling Stones.

There's a modicum of truth in this: Premier League footballers and pop stars do turn up at these shoots, but Man United lost to Swansea on Saturday and then to Sunderland last night, so I expect they've got other things than pheasant shooting on their minds; and from recent photos I've seen of the Rolling Stones you are more likely to see them at the Age Concern lunches at the pub, which are on Thursdays. To give some idea of the money involved, they, whoever they are, have probably booked a 500-bird day, that means they will have to pay for 500 birds whether they shoot them or not. But if they shoot more than 500 the excess will go on the bill! It could be £40 a bird plus VAT which is £48 a bird. It could be *more* than £40 a bird, I don't know. With tips we are talking £25,000 for a day out for eight guns.

The weather for them is settled. It's settled into a series of high winds and lashing rainstorms that turn up about twice an hour. I rent some buildings on the estate where they are shooting and when I go to feed my heifers later on, I can see the guns lined up outside a wood about half a mile away on the other side of a valley. I pause to watch what is going on but another squall arrives and the view is obliterated by the lashing rain. As I move on I must admit, as a man of the people, that there is a little bit of me that likes to see such wealth, getting wet. Years ago, when I was more important than I think I am now, I had a role in the dairy industry that used to get me invitations to corporate events including shoots.

The first time it happened I was invited to a shoot at a castle by a company that manufactured the stainless steel kit that goes into a creamery. I wasn't really sure what the 'form' was so I went down to the cash point and got out £100. We stayed in the castle the night before and at breakfast next day we had a sweepstake on the likely number of birds we would shoot. The cost £20 each. After breakfast we went outside and I was introduced to a retired police inspector who already had my gun under his arm and was carrying my cartridges. He loaded my gun for me all day.

At lunch we were told what we had shot thus far and were invited once again to guess what the final bag would amount to. Another £20 goes. In the afternoon I was starting to get concerned by my dwindling resources. There was a captain of the dairy industry shooting next to me who I got on very well with and during a lull I strolled nonchalantly across to him and asked him how much I was expected to tip the loader. £60, he said. Panic starts to set in and I suspect I will have to ask him to lend me some money. 'Well how much do I have to tip the head keeper?' 'Don't worry, the host pays for that.'

## FEBRUARY 1ST 2014

I've been thinking about the flooding and in particular I've been thinking about the unfortunate people whose homes and businesses have been ruined by flood water. Like a lot of people, I've been wondering why it happened? And as so often, the answer is under my nose, or out of your front window anyway. We live in a small river valley, the river is small, the valley is quite wide. It's flat bottomed and on the three or four miles of it where I live, there is not much gradient. So when we get a lot of rain, it floods. I can see flooded fields out of the window now. And it always has flooded, but over the last 20 years there is a difference to the flooding. It goes higher up the valley each time, there are hollows in adjoining fields that 20 years ago were ploughed and grew crops. They are now no-go areas which only grow rushes. So that is a clear indicator that the water table is now higher than it used to be, otherwise the rushes wouldn't flourish.

An even clearer indicator is that as the wet ground spreads higher up the valley an area of about five acres in the next field up has been abandoned for farming, fenced off and left to its own devices. The farmer concerned will be paid to allow this to happen, which is just as well, because if he didn't want it to happen, it would happen anyway – he would have no choice in the matter.

What has happened out there, three fields away from my front window, is symptomatic of the point I am about to make. None of us should be under any illusion: the countryside is ruled by environmentalists and wildlife groups. They tell the government, the government only has an eye on the ballot box, and they soon work out that there are more people in these groups than there are farmers. The government has its hands firmly on the farmer's purse strings, and they tell the farmers what to do. Make no mistake, there's nothing these nature groups like more than a flooded field. If I were to fence off half a field and let a stream turn it into a wetland, they would be in ecstasies. They would pay me well for allowing it to happen. I see these people often, parked on the side of the road, window down, binoculars at the ready, looking at a flooded area for the sight of a cormorant or a seagull or other inland water bird (thinly veiled sarcasm). Then it's off home to a nice warm house whilst somewhere else people are mopping up the mess in their kitchen.

I've often been puzzled by the phenomenon of the holy cows in India, who are so revered that they can stroll down the street with impunity and browse on the vegetables on market stalls. I never thought it would happen but there is a holy cow in the UK now, it's called the fresh water mussel. So determined are environmentalists to preserve this mussel and to see it flourish that it is the main reason that 'they' will not dredge silt from riverbeds, which is one of the main reasons there is so much flooding. If you google 'fresh water mussels' you will find that it has the most precarious lifecycle you could imagine but its protection is more important than you and your ground-floor flood. Local farmers went to a meeting where they were told lots of do's and don'ts, (mostly don'ts) about environmental practices.

I didn't go, I had a full report in the pub later. They were told they couldn't spread manure (it's not called manure around here) onto lying snow. Any farmer will tell you that it's quite handy spreading onto snow because the snow cushions the passage of machinery and you don't damage the field. Mustn't do that, the manure might get carried by the melting snow into water courses. So if there's a prolonged cold spell, come a thaw, where do they think all that salt ends up? In rivers of course. After the cold spell of two years ago they reckon there was so much salt going down the river Severn that the children were catching mackerel as high up the river as Shrewsbury. They reckon if you want to see a really filthy water course you need to look no further than where the drains empty off a motorway. Black with rubber and diesel – but that's OK. A farmer wouldn't dare do it. Like the man said in the story, it's their water, all of it, unless it floods your field or your factory or your kitchen, and then it's all yours.

### February 8th 2014

I've been thinking about time and our headlong rush to *save* time, and wondering what do we do with the time we think we have saved? All this has been brought on by reading a piece somewhere about the HS2 railway. I'm in a sort of dilemma about this. It's provoked a furore of mixed opinions, especially about its impact on people who live close to its proposed route. But if someone had just proposed building the M5, there would be the same reaction and we can't imagine life without that! I'm very interested in HS2's proposed time-saving. By its very nature, it can't stop at many stations, because all that stopping and starting would negate the high speed it travels at. So if the journey from the Midlands to London is cut by half an hour but you have to drive 45 minutes to catch the train, where is the saving? And what do you do with the time you think you've saved? Half an hour is no big deal – why not get up half an hour earlier? Why not work out what benefits you would get if you spent all that money on something else? Personally, I think we all should chill out a bit, slow life down.

\*\*\*

It was a part of the lot of the dairy farmer, because of the nature of his work, that he should only ever see half of a film or a rugby match on television. If something was on that he wanted to watch on a Sunday afternoon, for example, he would only see half of it because it would soon be time to leave the house to go out to milk. Similarly in the evening, who wants to watch a late night film who has to be up at four in the morning to attend to cows? In such a lifestyle, sleep becomes a precious commodity. So I've spent most of my life thus and sometimes the best you could hope for was that you would see the half you had missed on another occasion, even if as often happened, you had to put up with seeing the second half first, if you follow me. I enjoy good films but not enough to drag myself 25 miles to the nearest cinema. I've not been to the cinema for 30-odd years. I work on the principal that, no matter how big the block buster, it will be on TV before long, even if you only see a half of it!

All that is changed now because we have the technological ability to pause things and record things and satellite TV gives us choices we never had before. So most days I scan the movie channels to see what's on and probably record two films a week, that I can watch at my leisure. I might only watch a bit at a time, but at least the bits are in the right order.

This week I recorded Alfred Hitchcock's thriller, *The Birds*. Special effects have moved on a long way since that particular film was made, still very scary stuff. So we park films for a moment and move on to grandchildren.

Having grandchildren is one of the very best things that happens to you in life. In my various writings I've always written about them, to the point that once a good friend of mine actually told me that I mentioned them too much and they were no big deal. Years later he apologised and said I was right all along and it was one of the best things that had ever happened to him.

So last night we had my daughter's children here for the evening. The girl is 13 going on 23 and very sophisticated with it, the boy is 7, full of life and pesters you endlessly, mostly in a nice way. So she's trying to do her homework, and he's trying to disrupt it, and it's been going on, this disruption, for about an hour, stealing pens and nudging elbows and the like, tempers are getting frayed and violence is imminent and 'Leave her alone, David' isn't working any more. 'I'll quieten him down,' I say. I get the remote control and put *The Birds* on. I have his attention immediately, one seagull down the chimney and he's behind the settee, which is quite a good place for a seven-year-old boy intent on winding up his sister. Perhaps a bit cruel of me, but very effective. Roll on light evenings when I can take him on adventures.

## FEBRUARY 15TH 2014

The landlord of our pub is a Scot. He's the only Scot in there and we have to humour him – well, it's his pub isn't it? He goes on endlessly about Scotland and all things Scottish. All we get is 'Andy Murray', all summer, then we've had three months of 'Scotland will win the six Nations rugby this year', yeah right. I'm actually taking him to see Wales v Scotland later on. On my terms. I shall want to see what he's wearing before I let him in to the car. Who wants to drive all the way to Cardiff with his bony knees poking out from under a kilt? No one, believe me. But Scotland is ramped up even higher come Burns night.

I've been to Burns nights and I've always enjoyed them. I enjoy the poetry of the man, the eloquence and the perception of life in his writing. So we have our Burns night in the pub, not a huge number of us, but it was really enjoyable, the food was excellent and the company good. Most people made a contribution around the table: we had some Burns, some music on trombone and clarinet, some mouth organ and some stories. Oh, and plenty of Scotch whisky!

Years ago we always used to have a Burns night at the rugby club. Quite expensive to put on because we always had a piper and there aren't a lot of pipes around here. So we always had to sell a lot of tickets to make it work. Selling a lot of tickets was not easy and it took a lot of persuasion sometimes. We used to have Scottish dancing of the sort where you start off with one partner and progress on to others. I remember one young lad, who I had to work hard on to attend, saying, 'I'm glad you persuaded me to come now, you can start off dancing with your missus and end up with something quite smart.'

***

Belatedly, very belatedly, a Christmas story drifts in. I'm not sure it is a suitable story for a family audience but as I have absolutely no idea if anyone ever reads what I write, there's no way of knowing how old any readers might be, and anyway, I've seen a lot worse on television before the nine o'clock watershed. And I've always told you what I think needs telling, so here we go.

It's a very frosty Christmas Eve and there's midnight mass at the main church in a small town not far from here. It just so happens that the church roof is surrounded by scaffolding. Because there is a lot of lead up there, and that's where they want it to stay, the lead is both vulnerable and protected. There are security cameras everywhere and access to the scaffolding is blocked off. So the service is proceeding as normal when the choir master becomes aware of noises coming from the scaffolding. Rhythmic noises. The service comes to a slow halt and the vicar and choir master set off, fully robed, to investigate, accompanied by a churchwarden. As they could quite easily assume that they were about to tackle lead thieves, I think it is only fair to them to say that they went off to investigate, boldly. But it wasn't lead thieves they found. What they found was an inebriated young couple, stark naked despite the sub-zero temperature, perched precariously 80 feet above ground, in a very popular horizontal position, doing what young couples often do when they find themselves in those circumstances. It all gives new meaning to 'Ding dong merrily on high' doesn't it?

## FEBRUARY 22ND 2014

The news of flooding has saturated the news recently. Who knows what the situation will be soon, but we mustn't forget the part played in all this flooding by wildlife trusts and bird societies. They love nothing more than flooded fields and reed beds along river margins and have been lobbying for this for years. Now we can see where all this has led us. I was down at the rugby club yesterday, no match, the pitch was flooded, but there was a lunch. There was an egret out on the grass feeding and I thought how nice it was to see it, never seen one before around here. But we need a bit of perspective, a bit of balance. The pendulum has swung too far in the direction of these lobby groups: it's high time it was pulled back a bit, it's time they had their come come-uppance. The pendulum is best somewhere around the middle, it's where a balance is to be found.

\*\*\*

Our TB situation, recently, has a sort of 'limbo' feel to it. It's been put on ice, side-lined. It's true that we have calves everywhere, and we have put them everywhere you could reasonably put a calf, apart from in the farmhouse, but because we are in between calving batches, there are not any calves being born at present. So the only constant reminder of our TB-afflicted status has been the purchased milk that we have in the fridge, having decided that this was a prudent thing to do whilst TB was lurking about in our herd.

But limbo, like ice, doesn't last forever and our 60-day test is due. My emotions on this have gone from optimistic to pessimistic and back so many times that I now approach the test with a sort of resignation, as if it is now a regular part of our lives. So we do the test and we pass! I'm not sure I can believe it. At the previous test we had one animal they called 'inconclusive', so she had been kept in isolation and as she passed this time as well, we are free and clear at last.

But it's not all good news. My son-in-law rears beef cattle to sell on. He had 30 tested and five of the 30 failed so now he can't sell the 25 clear ones, only the five that failed will go for slaughter. This is a body-blow to him: the sale of these cattle had been critical to his spring cashflow. Spring is an expensive time of year on a farm, feed and fertiliser to buy, rent to pay.

***

Even in a normal winter, we don't get around our land much. As this winter is far, far from normal, you can imagine that getting around the land happens a lot less. It's been so wet that we've not been able to let the dog out for a run because he was making so much mess! It's been so wet you take a tractor or truck beyond a gateway at your peril and it's been no fun walking about, so getting around the land just hasn't happened.

But I did have occasion to get out of the truck one day, just to look over a gate at some grass we had sowed last autumn. And my eye didn't go to the grass. It was to three dead hares laid out quite neatly inside the gate. I phone the Keeper, he needs to know this, it's a part of our joint 'keeping an eye on things', knowing who is about. He tells me that he's seen two other little piles of dead hares elsewhere on the estate, he says there is hare coursing going on again and if the hare coursing is successful ('successful' being a term you would use according to your point of view), the coursers always leave the hares in a gateway where they know they will be found. I'm always trying to fathom out where people are coming from, what motivates them. If I were coursing hares, which is illegal, and if I killed one, it would be hidden in the bottom of a hedge. To put dead hares where you know they will be found can only be the hare coursers' two-fingered salute to all and sundry.

## MARCH 8TH 2014

I told you weeks ago the environmentalists were not without blame for the flooding. Quote in our Sunday paper 'We pay £3.6 billion a year to farmers for the privilege of having our wildlife exterminated, our hills grazed bare, our rivers polluted and our sitting rooms flooded.' Never mind that 60 million people go out every day on these islands looking for something to eat. Never mind that if you take good land out of production in this country, someone, somewhere, will cut down the same area of rainforest to make up the deficit. Never mind that they have to put that food on a plane to get it here.

***

I go to the pub on Thursday night, and for the first time since Christmas, it's a proper Thursday night, there's a good crowd there. So the farmers are discussing flooding in Somerset. 'One good thing, I bet there's not many moles left there.' This is brought into the conversation on the principle that it is an ill wind that doesn't blow some good somewhere, the other principle being that farmers don't like moles. For a moment I toy, in my mind, with the possibility of my lawn flooding, my lawn being a sort of mole sanctuary and the lawn-mowing season being imminent. But toying with an idea is a distraction I shouldn't have indulged in because the conversation has moved on.

   The next bit I hear is 'good things for warts, moles'. The speaker, conscious that he has everyone's attention, moves on. 'The master of the hounds phoned me up once and said his horse had a wart on its back and could I catch him a live mole?' 'What did he want a live mole for?' 'You have to have fresh blood from a mole to put on the wart.' Everyone silently digests this information. I'm sure they all have questions to ask but they are afraid to in case it shows a lack of knowledge that everyone else knows well. 'Not much blood in a mole,' he continues. How the hell did he know that? This is heady stuff, it's difficult to take it all in so we could do with a diversion to give us a bit of time. And here it comes. 'How do you catch a live mole?' 'Oh that's easy. You find a good run, you dig a hole below it, and you bury one of those old-fashioned sweet jars in it. Then the mole goes down the run and falls into the jar, simple.' 'So what did you do then?' 'Well, I take the mole in the jar to the hunt kennels. I kill the mole'. We didn't go into the detail of this so I can spare you that. 'Then, quick as you can, you stick a pin in the end of its nose so some blood comes out and while it is still warm, you rub it on the wart, not much blood in a mole, so you have to do it quick.' He settles back and start to roll a fag, conscious and pleased that he has had everyone's attention.

It's only me who wants the final detail. 'What happened to the wart?' 'Oh that fell off after just two weeks.' There's no capping that story. Someone says that the best thing for a wart is to rub it with some raw meat and then bury the meat up the garden, but this is small stuff compared with the mole story and besides, we've all heard that old wives tale before. 'I tried it once but it didn't work,' says someone. 'The dog saw me bury the meat.' I drive home from the pub quite content, two glasses of rosé inside me, and pondering moles and warts. Of course I don't believe any of this superstitious nonsense. If I get warts on a cow's teats, which happens occasionally, I tell a local witch I know and she charms them off. Then I buy her a gin and tonic. Never fails. And it's cheaper than raw meat.

**MARCH 15TH 2014**

On first appearances, it seems that things are progressing well on the farm, we would seem to be well up with our work. But it's only a perception. Because there's a lack of urgency in our daily routine at what is usually the busiest time of our year. There's a simple reason for all this. We are not busy at work because we can't do any. We can't take a tractor through a gateway. In a normal year we would have spread poultry manure on all our silage ground. We would have put fertiliser on all our grazing ground and our winter corn. The cows would be out at grass, by day at least. None of this has been done. It will be done eventually, it always is. But at some time there will be a lot of long hours involved as we play 'catch-up'. The implications of the wet weather will be with us for months to come.

When we do eventually get going on some field work, I look forward to it immensely. Up and down all those fields on the tractor. A chance to scrutinise every bit of every field and every fence, and a chance to scrutinise the wildlife.

I was doing some scrutinising of a different sort yesterday. I managed to mow the lawns. It was an excellent start because for the third year running, the mower started first time, always a good omen. But riding around on the mower is a bit like riding around on a tractor: you see everything, warts and all. And boy are there some 'warts' in our garden! There's a lot of wall been taken down/ fallen down, and all the things you didn't want people to see, that you hid behind the wall, are now in full view.

So I finished mowing on a bit of a 'low', conscious of all the tidying up still to do. I tried to cheer myself up by telling them at the pub that I'd mowed the lawns and that created a bit of a stir, first one to do it. Emboldened by the reaction, I told them I would have some new potatoes ready to dig for Sunday lunch. Didn't work the same.

\*\*\*

It's the end of February before the Canada geese turned up to nest on the island on our pond. Last year, the goose was driven off her eggs by red kites. I do not hold out much hope for her eggs this year either. There are kites everywhere; the skies are full of birds, but mostly kites, ravens and carrion crows. Some people have a view that there are too many Canada geese about, dirty, noisy, things. But the serious point has to be that if raptors can drive a large aggressive bird like a Canada goose off its nest and eggs, what do they do to other more vulnerable species? Only time will tell us the answer to that question. I fear the worst. But no one that could do anything about it seems to listen to points of view like mine. People have been banging on for years about man needing to 'manage' the man-made environment that is the Somerset Levels. No one listened to them either.

***

Have you noticed that there are not as many early spring lambs about as there used to be years ago? I can remember often seeing some lambs about in December and January. Come February and there would be lots about. I've seen lambs this year, but not many. It's all to do with saving costs. If your lambs are born in March and April they will grow as the spring grass grows. They won't need expensive cereals and expensive imported protein. They will be more natural.

It might not have crossed the sheep farmers' minds, but this is a more sustainable form of production. Grass probably always will be the cheapest food you can ever provide for a herbivore. And as grass declines in feeding value in the autumn, some of these lambs will move onto fodder crops, crops with a higher feed value than autumn grass, root crops like kale and turnips and rape. This is a modern equivalent of the old practices of folding root crops with flocks of sheep. It's not done with wooden hurdles any more, it's done with electric fences, but the effect is the same. Fertility is returned to the land by the manure from the sheep, the crop is harvested by the sheep, not by expensive machines and all the costs associated with that. It's a win-win scenario.

It's sustainable. From birth to abattoir, the sheep have eaten only what you can grow for them on your own farm. But you mustn't get carried away with all those 'win-wins'. Because every now and again, nature will send along a wet winter like we've just had, and if she does, a root crop is not the very best place to be. There's still a lot of last year's lambs on root crops and they have lost condition because of all the mud underfoot and the constant rain on their backs. So you can make your plans, and they will be fine on an average year, they will be excellent on a good year, but on bad years, they will be tested. Farming was ever thus.

---

## Also published by Merlin Unwin Books

**How Now?** Roger Evans £14.99 hb

**Pull the Other One** Roger Evans £12 hb

**Fifty Bales of Hay** Roger Evans £12 hb

**A Farmer's Lot** Roger Evans £12 pb

**Much Ado About Mutton** Bob Kennard £20 hb

**A Most Rare Vision** Shropshire from the Air Mark Sisson £14.99 hb

**Myddle** The Life and Times of a Shropshire Farmworker's Daughter Helen Ebrey £12

**How the Other Half Lived** Ludlow's Working Classes 1850-1960 Derek Beattie £14.99

**Extraordinary Villages** Tony Francis £14.99 hb

**The Countryman's Bedside Book** BB £18.95 hb

**A Job for all Seasons** My Small Country Living Phyllida Barstow £14.99 hb

**My Animals and Other Family** Phyllida Barstow £16.99 hb

**The Byerley Turk** The True Story of the First Thoroughbred Jeremy James £20 hb

**Maynard:** The Adventures of a Bacon Curer Maynard Davies £9.99 hb

**Maynard:** The Secrets of a Bacon Curer Maynard Davies £9.99 hb

**The Way of a Countryman** Ian Niall £16.99 hb

**The BASC Game Shooter's Pocket Guide** Michael Brook £7.99 pb

**Don't Worry he Doesn't Bite** Liam Mulvin £12 hb

---

Available from all good bookshops
For more details of these books: www.merlinunwin.co.uk